The Botanizers

Amateur
Scientists
in
Nineteenth-
Century
America

THE Botan

izers

Elizabeth B. Keeney

The University of North Carolina Press

Chapel Hill & London

© 1992 The University of North Carolina Press

All rights reserved

Manufactured in the United States of America

The paper in this book meets the guidelines for permanence and

durability of the Committee on Production Guidelines for Book

Longevity of the Council on Library Resources.

96 95 94 93 92   5 4 3 2 1

Library of Congress Cataloging-in-Publication Data

Keeney, Elizabeth.

   The botanizers : amateur scientists in nineteenth-century

America / by Elizabeth B. Keeney.

   p.   cm.

   Includes bibliographical references and index.

   ISBN 0-8078-2046-6 (cloth : alk. paper).

   1. Botanizers—United States—History—19th century.

   2. Botany—United States—History—19th century.   I. Title.

   QK46.5.B66K44   1992

   581'.0973'09034—dc20   92-5022

"Botanizing" (Frontispiece from William Whitman Bailey,

Botanizing: A Guide to Field Collecting and Herbarium Work,

1899)

In memory of Will Humphreys:

teacher, mentor, and friend

Contents

Illustrations

Acknowledgments

Like most books, this one owes its completion to many people. In the course of its evolution it has been touched—directly and indirectly—by an extraordinary number of hands.

Special thanks are due to Ronald L. Numbers who guided me toward the topic. My first crack at writing on the subject occurred in Daniel Rodgers's graduate seminar at the University of Wisconsin, where he and my fellow students provided invaluable feedback at a critical stage. Nathan Reingold and the Smithsonian Institution provided a year of research and thinking that enriched the project immensely. Carl Kaestle, the late William Coleman, Margaret Rossiter, Michele Aldrich, Barbara Melosh, Sally Gregory Kohlstedt, and David Allen all read and commented helpfully on part or all of one or more versions. Mark Barrow provided especially helpful comments and reassurance in the final stages. A. Hunter Dupree treated me like a colleague long before I had earned that, for which I am most appreciative. My colleagues in the History of Science Department at Harvard, most especially Everett Mendelsohn, listened patiently to pieces of this work.

During the revision of this work a number of friends provided invaluable moral support. Fred Jewett, Hank Moses, Jeff Wolcowitz, and my fellow Allston Burr Senior Tutors made my work days (and nights and weekends) as enriching as our students made them exciting. Susan Repetto, Dianne Weinstein, Catalina Arboleda, JoAnne Brown, Dwight Reynolds, Michael Engh, S.J., and John Warner all provided encouragement at critical points. The Usual Suspects Book Group (especially Karen Heath) nourished my mind and restored my spirits on a regular basis. My new friends and colleagues at Kenyon College have borne my obsessive discussion of revision during my first months here with remarkable tolerance and good humor. Lewis Bateman and Pamela Upton at the University of North Carolina Press have been extremely helpful.

Finally, I want to acknowledge a special debt. Because I am learning disabled, I am particularly grateful for having had more than my fair share of very good teachers. Will Humphreys, to whose memory this book is dedicated, was my teacher at The Evergreen State College before he became my mentor and friend. Will encouraged my interest in the history of science. More important, he believed in me so strongly that I began to believe in myself, without which no degree of compensation and mastery would have been worth much.

The Botanizers

Introduction

After rising to fashion during the 1820s, botany rapidly became the most popular science in America for recreational and pedagogical purposes, and it remained such throughout the century. Tens of thousands of enthusiasts embraced botanizing by collecting, identifying, and preserving specimens. Clubs, correspondence networks, specimen exchanges, and specialized publications arose to meet the demand for botanical culture. Young and old, rural and urban, male and female, joined together in pursuit of the natural history of plants.

The pathways through which specimens and information flowed included beginner and expert alike. While an increasingly influential few made science a lifework, the vast majority of members of the botanical community used science as a pastime. Without formal scientific training or employment, and with a deeper commitment to self-improvement than to the advancement of science, these botanizers, as they called themselves, had priorities and interests that differed from those of the nascent professionals. Yet while these differences led to disagreements over systems of classification and scientific standards, the botanical community managed to maintain a spirit of cooperation and collegiality for most of the nineteenth century. By the beginning of the twentieth century, however, the botanizers' focus on self-improvement had become incompatible with the professionals' interest in advancing science and their desire or need to seize autonomy and authority. Botanizers and professionals had drifted so far apart that neither group any longer considered botanizers to be members of the scientific community.

Despite the obvious importance of these changes, we know very little about the process of professionalization in general, and we know far less about the case of botany in particular, or the individuals involved in it. The accounts we do have view the change from the perspective of those who remained within the fold—that is, the professionals. This study takes a different frame of reference, that of those who do not fit today's conception of scientists: namely, the amateurs. The change can be readily illustrated by contrasting two

Vermont botanizers, one from the nineteenth century and the other from the present.

Cyrus Guernsey Pringle (1838–1911) took up botanizing in the wake of the Civil War in an attempt to restore his health, which had suffered from his imprisonment as a conscientious objector. Raised on a farm, Pringle derived his income from the culture and breeding of plants (especially flowering bulbs), the techniques of which he learned by experience. As a sideline, he collected botanical specimens for America's premier botanist, Asa Gray, and others, occasionally receiving payment. In the process he amassed a personal collection of approximately 50,000 specimens, which would become the nucleus of the University of Vermont herbarium. Completely self-trained and employed only on the fringes of botany, he was nonetheless a respected and valued colleague of the leading botanists of the country.[1]

The case of Henry Potter, a present-day Vermonter, is somewhat different. In July 1985 there appeared in the *Boston Globe* an article about Potter, then ninety-three, whose personal herbarium includes more than a thousand species, and who is arguably Vermont's "most knowledgeable resident on ferns." While still a child, Potter taught himself botany, spending winter afternoons in a haymow identifying the dried plants he found there with a field guide. Today, botanists from throughout the region occasionally consult Potter and eagerly await his upcoming book on ferns. Although his knowledge of ferns is self-taught, his doctorate is honorary, and he has never held scientific employment, he is, nonetheless, an authority in a small niche of science. Yet because he is a farmer with a ninth-grade education, no one—neither Potter nor those who respect his expertise—considers him a scientist or a botanist; rather, he is dubbed a "devotee" and a "pteridophile," or lover of ferns.[2]

Henry Potter's fame and the interest behind the *Globe* article are due in large part to the novelty, in this final quarter of the twentieth century, of an amateur who is an expert. Happy to consult him and use his expertise, professional botanists regard him as a resource, not a colleague. A century ago Potter would have been, like Pringle, a member—indeed, a distinguished member—of the botanical community. Today, however, he is a devotee, a pteridophile, not a botanist. This is largely an American phenomenon.

British amateurs, for example, are still integrated into their scientific communities to a great extent; in America, by contrast, integration is the exception, most noticeable in the astronomical and ornithological communities. Understanding the emergence of a botanical community, indeed an entire scientific community, composed exclusively of professionals is of fundamental importance to any examination of the structure and organization of science, of what it means to be a scientist, and of the place of science in modern American society. How it came to be that Cyrus Guernsey Pringle was a scientist, but Henry Potter is not, is the central question that I hope to begin to answer in this book.

Over the past three decades, scholars examining American science have devoted perhaps more of their energy to professionalization than to any other trend or development. Historians and sociologists have struggled to define "professional" in terms meaningful to the nineteenth-century context, to determine who those first professionals were, and to study the process of professionalization itself. Despite occasional rhetorical accusations that the issue is meaningless, this interest in professionals and professionalization has stemmed from a desire to understand a radical change in who did science and why, as the scientific community shifted from a predominantly amateur to a predominantly professional identity.[3]

The portrait of the amateur found in traditional historical studies has usually been painted from the perspective of the professional. Whether they have seen the amateur as a scientific enthusiast, a supporter of science, or a second-rate scientist, historians have understandably begun with the worldview of the professional because they were studying professionalization, and, perhaps, because they thought of themselves as professionals. As a result they have often overlooked the nearly self-evident fact that amateurs and professionals pursue science for different reasons. At its simplest, the mission of a professional scientist is to advance science and to further scientific knowledge. Many professionals, and thus many historians, have assumed that amateurs share this objective—an assumption most clearly reflected in the plethora of articles on the "Contributions of the Amateur," which overlook the relatively low priority that most amateurs attach to contributing to science.[4] Amateur botanists of the nineteenth century were far more intent on personal enrichment than on advancing science. More than any other factor, this led to the declining scientific status of amateurs in the closing years of that century.

Much of the confusion about nineteenth-century science can be traced to the problems of vocabulary that result from using terms like "amateur," "professional," "botanizer," and even "natural historian" and "scientist," that have changed meaning in important ways as the activities and individuals they describe have evolved.[5] "Amateur" first entered English in the eighteenth century with the broad meaning of "one who loves" an activity. It did not acquire the purport of "one who cultivates an activity as a pastime" until the early nineteenth century;[6] the corollary imputation of inferior status to the pursuer of a pastime came still later, somewhere in mid-century.[7]

"Professional" is a still more complicated word because of the dual senses of its root "profess": on the one hand it means "to avow," and on the other, "to take vows." From this dichotomy has arisen the modern split in usage between "one who has skill and expertise" and "one who has status because of membership in a highly regarded line of employment."[8] Both of these two rather different meanings of the word "professional"—expert and worker—concern us here. In opposition to "amateur," "professional" may imply merely the receipt of financial compensation. Often, today, this has positive connotations, because of the assumption that one must be somewhat proficient at something in order to make a living at it, as in the case of a "professional athlete" or a "professional artist." The connotation can, however, be negative, as in the case of someone who is paid for doing something that may better be done for nonfinancial considerations (e.g., a "professional politician") or someone who devotes excessive time or energy to an activity (e.g., a "professional student"). The meaning of "professional" behind these uses has two elements: remuneration, and commitment. When "professional" is used in this sense it suggests the seriousness and dedication of paid employment.[9]

In contrast, when "professional" is used in *opposition* to mere employment it means much more, suggesting a practitioner of one of a few specific callings—traditionally the law, medicine, and the church—that claim to have higher standards than ordinary work. This particular meaning carries with it notions of special training, certification, and employment in a field with authority and autonomy. Herein lies the problem. When we say that science becomes a profession in the United States during the nineteenth century, to what extent are we referring to the ability to make a living at science, and to what extent are we talking about these higher standards? Which comes first, and how are they related? The answers to these questions are far from clear, despite a generation of discussion among scholars of American science.

Those who have studied professionalization have found the emergence of the expert in the nineteenth century to be an especially important flag to the historian that something was changing. Indeed, we do not see the word "expert" being used until the mid-nineteenth century, because it was not until then that a significant number of individuals were able to make a living at being knowledgeable.[10] The increasing specialization of labor and of learning led at this time to the emergence of individuals with special knowledge and skills that allowed them to preempt tasks or information and thereby to realize autonomy.[11]

Like other areas of learning in America, science began the nineteenth century in a decidedly preprofessional state. To identify when science became a profession requires that scholars establish a working definition of "profession," a task that has received much attention but no universally accepted solution. Indeed one scholar, frustrated by the vagueness of many proposals, argued that the most commonly accepted criterion can be used just as well to describe skilled labor.[12] In the specific case of science, as I will discuss below, scholars disagree over the necessary ingredients for the status leap to profession.[13] Regardless of the precise definition used, it is crucial to recognize that in America professional authority and autonomy emerged only slowly as the nineteenth century progressed. With professionalization came dramatic changes in the role of the scientist in society, all built on expertise, whether real or presumed.[14]

"Defining professionalization is," in the words of Nathan Reingold, "a thankless task."[15] This has not, however, kept scholars from trying.[16] For contemporary scientists, professional identity is measured by specialized education, scientific employment, and affiliation with appropriate learned societies; as Reingold has observed, however, these criteria are inappropriate for an era like the nineteenth century when there were few strictly scientific jobs or societies, and little formal scientific education was to be had. Prior to the Civil War, fewer than half of the individuals whom historians have identified as "scientists" had a college degree or were engaged in full-time scientific employment.[17] Limiting the definition of "professional" to those few who met the modern criteria reduces the pool to a very small number. More importantly, it imposes a division that even leaders of the scientific community did not make until late in the nineteenth century. The terms "amateur" and "professional," as commonly applied in modern usage, are simply not useful or accurate categories for the nineteenth century.

Using the alternative labels of "cultivator," "practitioner," and "researcher" to capture the motivations of individuals, Reingold painted a more accurate portrait of divisions within the nineteenth-century scientific community. This recognition of different motivations, an attempt to see professionalization from the perspective of the most ordinary members of the community rather than from that of the leaders, provided new insights into the dynamics of that community. Reingold's "cultivators"—members of learned culture whose livelihood did not depend on science—pursued science as a hobby for personal enrichment. In contrast, his "practitioners" (those whose employment required scientific competence but did not emphasize research) and his "researchers" (leading scientists employed in research capacities) aimed to advance and diffuse science. While Reingold's new categories have not been adopted, even by Reingold himself, his recognition of the role of motivation is of lasting importance.[18]

Indeed, motivation provides what seems to me to be the key litmus test for the botanical community of the nineteenth century. Professional botanists, on the one hand, were those who acted to advance or further science, sometimes enjoying scientific training, employment, and society affiliations. Amateurs, on the other hand, sought enrichment for themselves, and sometimes for their students, through botanizing and were not dependent on scientific employment for a livelihood. This definition reflects the understanding of the nineteenth-century botanical community that professionals and amateurs pursued botany for very different reasons. An 1825 literary article written for the general public, for example, stressed such features as natural theology, accessibility of specimens, exercise, and gentility.[19] Clearly these were not prime motivations for professionals. This article, and many others like it, never mentioned the advancement of scientific knowledge, the discovery of new species, or other themes central to professional interests. While it is certainly the case that some authors appealed to utility, natural theology, and moral values less because they believed in them than to promote the social role of the scientist, it is also the case that for many individuals the appeal was very real.[20] Amateurs remained committed to self-improvement, the study of God's creations, and the pleasant but worklike character of botanizing throughout the century. As late as 1899, William Whitman Bailey's *Botanizing* stressed the usefulness of botany to gardeners and physicians, the development of good observation and reasoning skills, the ease of study, and the beauty of the subjects. Amateurs and professionals alike recognized that their ambitions were different. What they and historians realized less often was the influence of those differences on the science that these professionals and amateurs pursued.

When I began to pose questions about amateurs more years ago than I care to remember, some well-meaning skeptics argued that interesting as these questions might be (and not everyone thought they *were* very interesting), they were unanswerable because there were no sources. One cause of the tendency of historians of science to focus on professionals is their dependence upon the published and unpublished writings of professionals. Not all amateurs recorded their activities and ideas, and when they did, their writings were less apt to be preserved in either printed or manuscript form. The thoughts of botanizers can, however, be found. Within the traditional sources there appears—when one looks for it—a surprising amount of material directly dealing with amateurs. The authors of textbooks (often amateurs themselves) almost always used their introductions to try to convince amateurs to pursue the science, and indeed an amateur wrote the best-selling botany texts of the nineteenth century. Society records and proceedings are too scant to yield a statistical picture of amateur participation, but are nonetheless rich in descriptive accounts. Some learned journals contain accounts of amateur activities, along with occasional letters and articles from amateurs.

There is also a vast array of commonly untapped sources, heavily laden with information on amateur science. Popular journals such as *Godey's Lady's Book* and the *Youth's Companion* were full of articles on science by amateurs, as were advice manuals and children's schoolbooks. Popular fiction included accounts of botanizing written by amateurs or observers of amateurs. Finally, those who were not sufficiently distinguished as amateur botanists to warrant biographical attention in that regard were sometimes accomplished in other areas and well known nonetheless.

Thus sources revealing the experience of amateurs are surprisingly rich and abundant. They are simply different from those conventionally used to examine the experience of professionals. Indeed, this difference in sources is the first big clue to how very different the worlds of amateurs and professionals were, and are.

1 Botanizing

Over the course of the nineteenth century the practice of amateur botanists changed very little; professional practice changed dramatically. This simple contrast proved an immense distancing agent. It is, therefore, of vital importance to begin with a feel for what it was that amateur botanists were doing, who they were, and how many they were, before turning to what happened to their place in the botanical community.

Amateur botanists of the nineteenth century, and indeed most of those who remain today, were cast firmly in the natural-history mold. Their interest lay in the collection, identification, preservation, and exchange of specimens. Few branched into studies of life cycles, or physiology, or other more biological issues, even when these became the vogue among professionals.

When Europeans first set foot in the New World one of their self-appointed tasks was to catalogue its natural wonders. Driven by both curiosity and economic motivations, explorers and colonists sought to recount the splendor of the North American flora. After this first flurry of awed attempts to describe, and concomitant attempts to transport live specimens to Europe, both settlers and explorers turned to a more careful, scientific task of collection and description, and sometimes to the preservation of specimens for later identification and exchange. The science of botany grew with the task. The discovery of the New World caused chaos in the world of natural history with the deluge of specimens, many of previously unknown species, that flooded into Europe. Ancient authorities knew approximately six hundred species of plants, and mid-fifteenth-century authorities added only several hundred more. By the mid-eighteenth century, however, the number had risen dramatically to between seven and ten thousand.[1] This explosion in the number of items to be categorized required improved systems of classification.

The resulting disorder was somewhat alleviated in the eighteenth century when Carl Linnaeus's artificial system of classification became popular throughout much of Europe and in the English-speaking colonies of North America. The beauty of Linnaeus's system was its simplicity. Flawed as nineteenth-century scientists would find it, it was embraced by eighteenth-century naturalists for the ability it gave them to impose order on a seeming chaos of specimens. It required no special skill or equipment and it allowed the rapid pigeonholing of specimens. This enabled the observer to get a name fast—an attribute that would ensure its popularity among amateurs and educators long after it had ceased to be on the cutting edge scientifically.[2]

Throughout the nineteenth century, amateurs practiced taxonomic botany almost exclusively. While the methods and details involved changed over the course of the century, the basic activities remained much the same: procuring, identifying, preserving, and exchanging specimens. Not every amateur pursued each phase of botanizing. Many only collected and identified specimens; others created extensive herbaria. Some concentrated on one family of plants or a small collecting area, while others collected more broadly. What unified their activities was a natural-history focus, which at first aligned them with the mainstream of professionals and then separated them from it.

The experience of collecting varied greatly depending on what, when, where, and who. At one extreme we have the harrowing description by William Bartram (1739–1823) of his nearly becoming an alligator's meal;[3] at the other extreme lies the garden-based botanizing of Jane Colden (1724–66) in New York.[4] What unites these individuals is mission: both collected specimens. In the case of Bartram, far from the conveniences of civilization, the aim was often to make a hasty field identification or create an adequate description with words and illustration for later confirmation, or to procure seeds or roots that would survive the journey back to Bartram's Philadelphia home or his London sponsor's garden. Jane Colden, in contrast, rarely strayed far from her New York home to procure her specimens, although she also worked on material provided by her father and correspondents. She was somewhat of a novelty, having mastered the Linnaean system despite her gender. Colden's greatest work was her paintings.[5]

By the 1820s the number of collectors was beginning to swell. Classes in botany—which would become an extremely important component of the broader lyceum movement—existed in America as early as the 1780s.[6] Such classes and lectures varied in the relative emphasis given to entertainment and to edification. By 1810 Amos Eaton had begun what would be a lifetime of popularizing botany.[7] The result of such popularization was the rise of botanizing, an activity that was less focused than William Bartram's expeditions and more scientific than Jane Colden's flower painting.

Who

Who botanized? This is a broad and difficult question to which there are no firm answers. There is, for example, no sure way to estimate the *number* of Americans who botanized in the nineteenth century. Even historians attempting to count only the more visible members of the community—those whose employment was in science, or who published extensively—have despaired at that task. Botanizers, many of whom never wrote anything, or made a cent from botany, or joined an institution, or subscribed to a botanically inclined periodical, are still more elusive. It is nevertheless possible to get a sense of the size of the population despite the lack of firm numbers.

Historian Nathan Reingold addressed the problem head-on in his classic article "Definitions and Speculations: The Professionalization of Science in America in the Nineteenth Century." Before one tries to count or estimate, it is essential to specify very clearly what one is counting. Confronted by the shifting lines between amateurs and professionals as the century progressed, the tangential nature of those lines at any time, and their anachronistic nature, Reingold abandoned the categories entirely; looking at the period 1800–1850, he portrayed a scientific community that ranged from the loftiest "researchers" to the commonest "cultivators" (members of learned culture who pursued science for personal enrichment). He estimated the entire scientific community for this period at 14,200, with 11,000 of those being cultivators.[8] Botany's popularity vis-à-vis other sciences was such that I would estimate that of those 11,000, one-third to one-half were botanizers.

These several thousand members of learned culture, however, by no means constituted the entire botanizing population: what of children, and for that matter women, who were not part of learned culture? While this number is certainly too low, the claimed sales figures of popular textbooks and field guides err in the opposite direction. Amos Eaton's *Manual of Botany* sold something on the order of 2,500 copies.[9] The two most popular botany books of the nineteenth century claimed to sell 270,000 and 375,000 copies each,[10] and another sold 100,000.[11] Literacy and several dollars, rather than learned culture, were all that was required to use these books. Just as it would be erroneous to assume that only members of learned culture were active botanizers, so it would be foolhardy to suggest that everyone who purchased a textbook or a field guide pursued the science for more than a few weeks, or even read the book through. Even a conservative estimate based on such figures, however, would appear to put the number of botanizers at several hundred thousand, rather than several thousand.

The same problems involved in estimating the number of botanizers also make it difficult to determine exactly *who* botanized. There are, however, some obvious clues. A botanizer needed, for example, rudimentary literacy in order to find the identity of a specimen in a field guide.[12] Also requisite were the economic means to spare at least a little time for an activity not essential for survival, and several dollars to purchase a field guide. Taken together, literacy and finances probably ruled out roughly the lower third of society, including virtually all blacks, for most of the century. Age and gender were minimally affected by these dual barriers. Many individuals began botanizing as children and continued into adulthood. Women participated enthusiastically, but whether in greater or lesser numbers than men is unclear.

Quantitative studies of American scientists often begin with a quandary: Who was and who was not a scientist? In other words, "Whom shall we count?" We have learned a great deal from studies that use as their sample individuals who published productively, appeared in the *Dictionary of American Biography* or the *Biographical Dictionary of American Science*, or were members of the American Association for the Advancement of Science. Shifts in class patterns, religion, education, and employment all emerge clearly. What does not emerge, however, is a sense of the changing size of the whole community, or of the community of amateurs, or of botanizers. [13]

Additional evidence about botanizers comes from the literature of popular botany. The arguments that writers presented when championing botanizing emphasized middle-class values. Self-improvement, the work ethic, and Protestant theology all pervaded the literature of botanizing. Narrowly interpreted, this tells the historian little more than that those who wrote about botanizing were set squarely in the middle of society. One can, however, cautiously surmise that the majority of authors knew for whom they wrote and were not hopelessly out of touch with their audience— the latter being, after all, largely white, middle-class, literate, and Protestant.

Where and When

The first step in botanizing was the procurement of specimens. This commonly meant going out and collecting. The casual botanizer might simply roam the neighborhood in search of plants or pick up specimens in the course of daily travels. Rural physicians, like S. B. Mead of Augusta, Illinois, and ministers, like Thomas Morang of Ashland, Massachusetts, found the opportunity to collect during their long travels to and from calls. [14]

More dedicated botanizers made special trips in search of material. Indeed, many collectors regarded walking in search of plants as an excellent source of exercise, gentle yet invigorating. One advocate described a sixteen-mile, cross-country collecting hike in the 1840s: "through narrow lanes, and past rustic cottages, with their lilacs and roses, their simple gardens, and their loquacious inhabitants; then down through woodland paths, and over meadows spangled with violets; through bogs and over potato-patches; scaling precipices and wading through ditches, slaking our thirst with the water of musical streams, and appeasing our hunger with a few scattering strawberries, made the whole day one of intense delight."[15]

While particularly enthusiastic botanizers sometimes took long collecting trips, most stayed fairly close to home. Quiet roadsides, woods, and meadows were ideal environments for collecting. One botanizer recounted a rural roadside walk that had yielded forty-three species.[16] Another pointed out that vacant lots in cities offered a profusion of "weeds."[17] Foot, horse, railroad, and eventually bicycle all transported botanizers. Foot offered the best opportunity for stopping to examine specimens. Horses traveled farther and faster but, as Benjamin Smith Barton discovered, sometimes ran off and left the collector stranded.[18] Railroads were more dependable (though not fail-safe), and some enthusiasts made a game of being able to recognize species from a rapidly moving train.[19] As public transportation spread out to suburbs, urban botanists found their ranges greatly expanded. One Illinois botanizer wrote in 1891:

The electric car lines have been extended in several directions, one or two miles beyond the city limits (mainly to boom suburban lots) and will afford me facilities in reaching some very desirable botanizing grounds which last summer were beyond my walking ability. One is the hills above the city on the bank of the Rock River which last summer I could only visit twice. Now I can go within a mile by the cars. The other is one upon which I am counting greatly, viz., the right-of-way of the Chicago & Northwestern R. W. which was fenced in thirty years ago and has of course never been cultivated or pastured since.[20]

When collecting, one needed to take care to get a complete specimen—that is, "one or more shoots, bearing the leaves, flowers, and fruit; and, in the case of herbaceous plants, a portion of the root."[21] Large fruits and seeds, and sections of wood, could be collected as well.

Botanizers collected in all seasons and at all times of day. Many botanizers were busiest during warm weather, when flowering plants were in bloom, but others studied trees or nonflowering plants all year long. Even those who specialized in flowering plants could occupy themselves out-of-season with preserved specimens or book-learning. Many botanizers collected catch-as-catch-can in the course of their routine daily activities. Those setting out specifically in search of specimens generally favored midday, when the dew had evaporated and light-sensitive flowers were open.[22]

How

At its simplest, of course, botanizing required no equipment and less fuss. Henry David Thoreau carried specimens in his hat, and not all collectors felt a need to dry or permanently preserve specimens.[23] A coat with big pockets (one veteran suggested a hunting jacket, with a large game pocket and many other small ones) could serve double duty as both outerwear and storage.[24] Others preferred to store specimens in a heavy bag or specialized boxes and presses. Those going after small or wet specimens found jars, vials, and envelopes helpful. A knife or trowel was useful for digging or cutting-free specimens, as well as for making sections.[25] Nets or dredges enabled collecting from pond and sea, though sometimes a boat was also needed. Pond lilies, for example, were sufficiently difficult to dislodge that even from a boat one risked falling in if care was not taken (though one could always save face by saying that this was a deliberate attempt to observe the underwater portion of the plant!).[26] A hand lens was invaluable for small work in the field and at home. At mid-century a suitable single lens, which folded into a case for carrying, sold for a dollar or less.[27] A simple dissecting scope sold for $1.50.[28]

Those unfamiliar with the region in which they would be collecting were advised to carry and use map and compass. One old hand recommended maps for all collectors so as to pinpoint the collection site. He suggested mounting the map on cloth, varnishing it, and then folding or cutting it to a convenient size; information about water sources and terrain could be written directly on the map for future reference.[29] Similarly, a notebook was regarded as invaluable. Serious botanizers used a notebook to record field notes and sketches, thus providing a permanent record of the day's activities; by using small, flexible books of uniform size, the botanizer could begin to compile a library of notes. One suggested model featured a small hole in the bottom of the spine through which a string was secured, attached to a pencil: the string could then serve as a bookmark, and a pencil was always handy.[30]

An essential tool for botanizing either at home or in the field was a flora or manual of the plants of the region. Finding a manual that was both detailed enough to be worth carrying, and small enough to bring along conveniently, was the bane of many botanizers. Asa Gray's comments to a fellow botanist urging him to publish a field guide were not unusual: "As to 'Pocket Manual of Botany,' the simplest solution is one which some of my botanical friends have made, viz, that of cutting a copy of my manual [?] close to the type, and binding in limp leather. Copies could be prepared in this way if there was a demand, and they could be printed on thin paper."[31]

William Whitman Bailey included instructions on using manuals effectively:

> If it is possible to obtain a pocket manual of the plants of the region in which you propose to collect, it will prove a valuable accessory, provided the time in the field is not too limited to permit its occasional use. When time is available for consulting it on doubtful points, the field work is much more profitable and fascinating, and a great amount of discrimination in collecting different species is possible even to the novice. In such a book all the plants which are known to occur in your region can easily be indicated by some appropriate symbol or colored ink. . . . It can also be used as a check-list for herbarium specimens, and in many other ways which convenience, necessity, or fancy may dictate.[32]

With manuals costing between $1.25 and $2.50 at mid-century, the expense of being a botanizer was certainly far less than that of being an amateur chemist, microscopist, astronomer, or even geologist.[33]

Depending on the length of the proposed journey, food and drink might need to be brought along. Whit Bailey recommended carry-

William Whitman Bailey (Courtesy of Hunt Institute for Botanical Documentation, Carnegie Melon University)

ing some light and portable food, such as a chocolate bar, for use in the event of an emergency. Water was to be carried in a canteen or flask, unless one could be absolutely certain of both the availability and the potability of water along the way.

Those heading on long trips needed to consider medical emergencies as well. Bailey's advice was concise: "In malarial districts, when the excursion is at least of several days' length, it is well to carry quinine in the form of pills, as a possible preventative of ague. A mild laxative is also desirable. To these we should add a small amount of brandy or whiskey as a stimulant in emergencies, and strong ammonia as an antidote to the bites of poisonous reptiles and the stings of insects."[34] To ward off obnoxious and potentially dangerous insects, he recommended protecting oneself by carrying a headnet and possibly a repellent of some sort.[35]

Clothing for botanizing depended on the gender of the collector and the type of collecting to be undertaken. Men and boys required nothing special for a walk through a yard or along a quiet roadside. The botanizer pictured in the frontispiece of William Whitman Bailey's *Botanizing* certainly fits the author's recommendation that the botanizer's clothes must be "comfortable and serviceable." Bailey occasionally took his own advice a bit too enthusiastically, in the eyes of his family, who were embarrassed when neighborhood children took him for a tramp.[36] Bailey was not the only botanizer whose dress distressed his family. Coe Finch Austin's wife begged him not to wear a suit she had just made for him when he went to visit a friend who lived near a favorite botanizing spot. Austin insisted he was simply going visiting, not collecting, but by the time he got home he was disheveled and had ripped out the lining of the jacket to wrap small specimens, while the coat itself was serving as a bag.[37] "One cannot," as Bailey commented, "lay down precise rules for anything so largely influenced by idiosyncracy [*sic*], caprice, or experience."[38] And yet, as he remarked, "The average botanist takes care that his clothes fit, that his shoes do not pinch, and that the fabric of his garments shall withstand the grasp of briars, come out well from impromptu baths, and allow him to climb, wade, run, or accommodate himself to any human vicissitude."[39] The fabric should be wool or corduroy, with a flannel undershirt, and a waterproof coat or suit was also suggested.

Shoes were of obvious import, needing both to fit well and to provide protection over difficult terrain. Hobnails were a help in mountainous regions. Rubber soles were not durable enough. Rubber boots, however, were desirable for wading, and high-top leather boots were a must in snake country. Extra socks were always carried on a journey of any length. Slippers for camp were worth taking on an overnight excursion.

Headgear varied with the situation. Brims and vents were desirable for protection from the sun; a close-fitting cap was in order when wind was a problem. An umbrella, while inconvenient, doubled as an implement for collecting fruits or insects shaken down from the branches of trees. A cane or staff, a useful tool for walking, could also be used to pull down an elusive branch or draw a pond lily closer to shore.[40]

For women and girls the issue of clothing was more complicated, since traditional female clothing was ill-suited for collecting. Botany was considered "lady-like" and hence "lady-like" apparel was in order—yet skirts caught on bushes and became soaked in damp grass, and tight bodices restricted movement. How to survive out-of-doors in full-length skirts and genteel shoes? In part, the answer to the quandary of what to wear came from differences between male and female behavior in the field: many female botanizers solved the dilemma of how to botanize and be genteel at the same time by restricting their collecting to very "tame" situations, or by recruiting males to collect for them.

Those who chose to do their own collecting, however, met the challenge in a variety of ways. Mary Treat, a Southerner, observed that "crinoline and long skirts are entirely out of place in woods and in climbing mountains."[41] The reform or "Bloomer" outfit was a solution for some. Others simply adapted what they had, or borrowed from male relatives. Surely Kate Furbish did not botanize in the formal full dress in which she is most often shown in portraits, when she collected the specimens she sent to Harvard with a note apologizing for their condition, saying: "I could not carry a vasculum up the mountain. It was a two miles' climb over rocks and fallen trees, and the air was warm and the black flies thick."[42]

There were alternatives to collecting one's own specimens. Collectors could be hired and often were, especially for botany classes. Women were not always free to collect and often worked from material procured by men and boys. Specimens could also be purchased. Periodicals often carried the advertisements of collectors who were willing to sell dried and live specimens for as little as five cents each plus postage.[43]

Once material was at hand the next step was identification. For most botanizers well into the century this meant using a key based on the Linnaean system of classification. For others, it generally meant using some artificial means to locate a species within either an artificial or a natural system. In the Linnaean system the first step was to observe the number and arrangement of floral parts, especially stamens and pistils; other keys began with floral color or the number and arrangement of petals. Whichever system was used, one progressively narrowed down the possibilities until only one, or a few, remained, and then confirmed the identification with a description. For most amateurs, ease of accurate use was the criterion for a good manual; the underlying system of classification mattered only in this context.

Once collected (or otherwise obtained) and preliminarily identified, specimens were either placed in a tin box or inserted into a press to be preserved through drying. Tin boxes were ideal for collecting material to be used soon, especially for classroom specimens. Botanical boxes specially designed for the purpose—vascula—were square or oval tubes, with a hinged lid running along the length, and fitted with a carrying strap. Indeed, such vascula became an emblem of the botanical community, serving as a means of mutual recognition. Many found that other metal boxes, especially those designed for candles or sandwiches, were adequate.[44]

For specimens to be studied more than a day or two later, drying was necessary. At its simplest the drying process involved placing a flower in a heavy book. One botanizer remembered pressing specimens in "the Congressional Reports and between the leaves of other books of equal interest to the average reader."[45] Special plant presses—two tightly bound boards or wiresheets on either side of layers of blotting paper—produced better specimens and accommodated thicker material.[46] Specimens could be placed directly into a portable press in the field. Portable presses varied greatly in size, but the consensus was that something on the order of ten by fifteen inches was about right for most sorts of specimens, though those specializing in smaller species needed less.[47] Presses were somewhat unwieldy, and many were fitted with either a shoulder strap or back straps so they could be worn like a knapsack. Plants were to be dried under strong pressure, with heat if desired. The paper was to be changed at least daily until the specimen was completely dry (2 days–1 week, depending); the sheets could be dried for reuse by punching a hole in each one and running a string through them in order to hang them between hooks, rather than "spreading them all over the room."[48] Delicate specimens were best placed between two sheets of very thin, absorbent paper, which could be nested in the dryer between successive fresh papers, thus eliminating the need to transfer small or fragile specimens from one sheet to another repeatedly.[49] At the time of collection, notes should be taken about the locale and date, and this information should remain with the specimen to avoid subsequent confusion.

Once dried, the specimen was attached to a thick sheet of white paper of a standard size (recommendations varied, but eleven and one-half by sixteen and one-half inches was a popular size)[50] and was securely labeled with information on the identification of the specimen and the specifics of collection (when, where, and by whom).[51] These easily mailed, flat sheets were exchanged with other collectors. They could also be arranged into genera and orders and either bound or stored flat in drawers or shelves.[52] So stored, a collector's herbarium could be studied at leisure in the "off-season."[53] Michael Schuck Bebb, a self-taught authority on the willows, did much of his work as a means of whiling away winter evenings: "I have been fussing over Willows, my usual winter occupation, and have really become quite enthusiastic. . . . It is only in this way, now, that I can keep up the pleasant illusion that I still belong to the fraternity of working botanists."[54] Even children were encouraged to arrange their specimens in an orderly fashion, so that they would learn the maximum amount about botany and also learn to think in an orderly fashion.[55]

The natural-history focus of botanizing, influenced by a host of concerns that had less to do with science than with social values, dominated amateur botanical activity for much of the century. While the practice described above closely resembles that of taxonomic botanists today, it bore less and less resemblance as the century progressed to what the leadership of professional botanists were doing. This would, as we shall see, have profound implications for the health of amateur botany in America.

2 Information Networks in the Botanical Community

Early-nineteenth-century Americans with an interest in botany enjoyed access to a wide variety of information sources. Societies, journals, and a handful of key individuals and institutions all facilitated interaction between amateurs and professionals. This interaction was crucial to botanizers, whose greatest need was information. For much of the century, the preprofessional state of the discipline meant that amateurs were well integrated into botanical networks. They participated in local, regional, and national societies that brought together individuals interested in higher learning, natural history, and botany itself—indeed, amateurs formed the majority in most botanical societies, and in many learned and scientific societies, until late in the century when the number of professionals reached the critical mass necessary for a "professionals-only" society to emerge.[1] Similarly, journals, which informed botanists of developments in their field far faster than could books, catered to broad audiences in order to survive financially.

Perhaps even more important than formal institutions, such as societies and journals, were the informal information and specimen exchange networks. Amateurs were the legs, hands, and eyes of individuals and institutions—notably John Torrey, Asa Gray, and the Smithsonian Institution—that served as great clearinghouses, linking amateur and professional botanists together in a national network for the exchange of specimens, information, and materials. Amateurs got advice, a wider range of specimens, and sometimes supplies or cash. In return, the clearinghouses received specimens from throughout the country, greatly broadening their base of knowledge without the time or expense of fieldwork.

The significance of these societies, journals, and exchange networks to the professional community is well documented.[2] During the half-century between the mid-1830s and mid-1880s that marked the heyday of the amateur botanist in America, cadres of willing disciples freed those at the pinnacle of the discipline from the time-consuming business of the fieldwork that the then-prevalent natural-history focus of botany required. This arrangement, which declined as botany became more professionalized and more biologically oriented, enabled Torrey, Gray, and others to produce the synthetic masterpieces that became their hallmarks.[3]

The importance of these same societies, journals, and exchange networks to amateurs, however, and the role that amateurs have played in them, have received less attention.[4] While it is certainly the case that we can enumerate the contributions of the amateurs, an emphasis on those contributions can lead us to overlook what the arrangement meant to the amateurs themselves. By examining the amateurs' end of networks, we can illuminate the experience of botanizing and the evolving relationship between amateurs and professionals as the two groups grew more distinct.

Until very late in the century, the number of professionals, no matter what definition we use, was minuscule. Somewhere between the obviously amateur individuals conversant with botany who occasionally collected a few plants, and the equally clearly professional scientists for whom botany was their lifework, lay a group not fully amateur *or* professional, distinguished by a proficiency with the local flora and by contact with others with whom they shared specimens and observations. More proficient and perhaps more dedicated than the average amateur, these individuals were nonetheless often self-taught, unable to derive a steady income from the pursuit of botany, and possessed of largely amateur values and goals. It would be easy to suppose that those amateurs who contributed regularly to the herbaria of the major leading botanists were closer to the professional end of the ideological spectrum because the fruits of their labor took the form of professional work—but this was by no means the usual case. Even for many amateurs of this dedicated and proficient level, improvement and pleasure were often the driving forces. Their contact with more professionally oriented botanists was not so much an attempt to aid science as a means of aiding themselves. Even though botanizing was a popular pursuit, the most dedicated sometimes found themselves isolated from others who understood their passion. If no one locally understood, perhaps a correspondent might.

Amateurs and professionals had much to offer each other, and thus the organized channels served both groups well. Often, professionals and amateurs alike profited from their mutual endeavors; indeed, they worked together amicably for much of the nineteenth century precisely because of the mutual scientific and social benefits they reaped. Amateurs gained sorely needed access to actual or relative expertise. Moreover, the esprit de corps fostered by societies and journals was just as important to amateurs as to professionals. All the members of a society saved money by collecting in groups and pooling libraries and equipment. All concerned benefited when a society or journal was enabled to survive by the financial support of amateurs. Early- and mid-nineteenth-century amateurs were not stifled by their interaction with professionals, but instead saw professionals as a source of expertise and guidance.[5]

The development of these organizations and networks exemplifies the changing relationships of amateur and professional botanists in nineteenth-century America. Prior to 1846, botanists of all levels of expertise and experience joined others interested in science as near equals, coexisting in societies, reading and writing for the same journals, and freely exchanging information. The creation of the Smithsonian Institution in 1846 and the American Association for the Advancement of Science in 1848 symbolized the dawn of a new era.[6] During the third quarter of the century professionals and amateurs often utilized common societies and journals, but professionals increasingly sought to guide the efforts of amateurs, whom they began to value more as aides than as colleagues, and they came to dominate decision-making and leadership positions.

The Early Period

At the dawn of the nineteenth century, the information flow within the scientific community reflected America's relatively egalitarian social structure. The colonists had created networks through which novices and experts supported each other and shared what meager resources there were. Few Americans were able to devote their full attention to spreading and diffusing knowledge of any sort, and fewer still to pursuing botany. The spirit of cooperation among intellectuals carried over as the new nation began to create its own institutions. The societies and journals of the early republic aimed to serve as broad an audience as possible. Breadth was financially expedient and logistically necessary, because of the small number of true devotees. Furthermore, the philosophical underpinnings of the new nation dictated egalitarianism and antielitism. There was, prior to the 1840s, no reason to consider creating an elitist scientific establishment, because there were few who would benefit.

Networks

The earliest networks, which emerged in the eighteenth century—notably the Royal Society of London, and the informal correspondence web known by historians as "the natural history circle"—were international in nature. While the Royal Society was selective about whom it elected to membership, it was always delighted to receive correspondence or specimens from the colonies. The most valued colonial correspondents, fifty-three of them by the start of the American Revolution, were elected "colonial fellows."[7]

Also based in London was the natural history circle, a network of colonists and Europeans who were the predecessors of the nineteenth-century clearinghouses. America played a special role in this network. The natural history of the colonies had captured the imaginations of European naturalists since their earliest contacts with the New World. By the 1730s an "international effort to advance natural history" had emerged. At the natural history circle's center was London merchant Peter Collinson, who used his business connections to establish new scientific ties to the colonies. Collinson received specimens and scientific correspondence, which he then passed along to the great European naturalists. Collinson's special interest was botany, and the circle reflected that passion. Among the colonists within the circle were John Bartram, James Logan, John Clayton, and Cadwallader Colden, all of whom devoted most or all of their scientific energy to botany. Each traveled and collected. Behind each of them was a smaller network of similarly inclined individuals, feeding their findings on up through the system; John Bartram, for example, received specimens from correspondents up and down the eastern seaboard.[8]

Societies

Colonists also laid the foundation for America's first major scientific society, the American Philosophical Society, which emerged in the 1760s from Ben Franklin's Junto. Though it professed to be a national society, most of its active members lived in or near Philadelphia. It was the sponsor and publisher of much botanical exploration. It reached amateurs through its publications and, in the case of the most dedicated and accomplished, through membership. While on a national level the American Philosophical Society was of limited help to amateurs, a few Philadelphia-area amateur botanists benefited by participating directly, using it as a local society—an institutional form that thrived in the early republic.[9]

Between 1815 and 1845 the number of learned and scientific societies grew by over 350 percent. If we count only those societies that included botany among their interests, the number rose from eight to thirty—a 275 percent increase. Cities as diverse as Providence, St. Louis, Baltimore, and Philadelphia could boast of active scientific societies by the mid-forties. Most were situated in New England and the Middle Atlantic states, where population was densest and communities were sufficiently established to consider expending resources on cultural activities.[10]

Scientific societies in the early republic vividly illustrate the necessity of a large membership pool for survival. For example, the Academy of Natural Sciences of Philadelphia, which was founded in 1812 and included botany among its interests, encouraged the participation of amateurs and novices. It offered popular botany lectures in 1814 and 1815 that drew "upwards of two hundred ladies, besides a considerable number of gentlemen" for the first series, and more for the second. Not only did this ensure the Academy's success, but it also put the older, more selective Linnaean Society out of business.[11] Similarly, the Providence Franklin Society was founded in the twenties by a group of men who wished "to mutually aid each other in the investigation of philosophical subjects." The members soon saw the benefits of broadening their scope to include "the propagation of a taste and love for the things and phenomena of nature among people." The Providence Franklin Society regularly invited the public to participate in its field trips, courses and lectures. By thus opening its doors to the public, it broadened its base of support and ensured stability.[12]

The consequences of an insufficient base of support are evident in the fate of the Charleston Botanical Society and Garden, established in 1805. Its founders were physicians and gentlemen-scholars who envisioned the Botanical Society as an adjunct to their local medical society. They were able to raise considerable interest among fellow physicians and scholars for the first few years, but the pool of potential members was too restricted to sustain sufficient support for the garden and society.[13] The causes of the Charleston society's failure are not clear. However, the most obvious difference between the successful Academy of Natural Sciences and the Providence Franklin Society, on the one hand, and the short-lived Charleston Botanical Society, on the other hand, was the latter's narrow base of support.

Journals

Like societies, journals of the early republic had to appeal to a broad audience in order to survive. With the exception of a few annual society proceedings, no American magazine survived as long as eight years until 1805. However, the number of journals grew rapidly in the early decades of the century, from twelve in 1800, to forty in 1810, to just under a hundred in 1825; by 1850 there were nearly six hundred. Many of these journals included at least some coverage of science.[14] Journals with botanical information came in as many wrappers as the societies and reached still

The Josselyn Botanical Society on an outing from Farmington, Maine, to Hillman Cascade near Industry, Maine, 9 July 1896. Kate Furbish (see chap. 5) is in the second wagon from the right, seated in the front seat to the right of the driver. In the field portrait, she is number 19. (Courtesy of the Archives of the Gray Herbarium, Harvard University)

more people. Societies were accessible primarily to urban dwellers because they required a concentrated population to provide enough interest, and the membership of most, though not all, was composed of white males; journals, in contrast, required only that one pay the subscription (between one and six dollars per year for most). From early in the nineteenth century, articles on botany appeared in an ever-growing array of journals, from the most popular, to those aimed at a more scholarly audience, to those dedicated to scientific affairs.[15]

The most prestigious scientific journal of the nineteenth century was the *American Journal of Science*, founded in 1818. It endured despite a circulation of only five hundred to one thousand, low even by nineteenth-century standards.[16] The case of Increase Allen Lapham of Milwaukee illustrates the importance of the *American Journal of Science* to the devoted amateur. Lapham, who pursued most areas of natural history at some point in his life, was especially fond of botany. He became a subscriber in 1827, at age sixteen, the year before his first article was published in the *Journal*, and he continued subscribing until his death, dropping the *Journal* only during a year of extreme financial difficulty. Indeed, his biographer uses that lapse to illustrate just how dire his circumstances were.[17]

The coverage of botany in the *American Journal of Science* dated from the journal's inception and continued throughout the century, with especially strong coverage from 1834 to 1888, the years in which Asa Gray wrote or edited much of the botanical content. The first issue of 1842, for example, devoted 56 of its 216 pages to Gray's lengthy account of a botanical trip to North Carolina, plus several pages of descriptions of Ohio plants, and 5 pages of botanical book reviews.[18] Reviews, original research, botanical notes, and news of activities made the *American Journal of Science* a valuable resource for Americans interested in botany, whether amateur or professional.[19]

Journals of less specialization but greater circulation were equally important to amateurs, indeed probably more important. Medical and agricultural journals, such as the *Medical Repository* and the *Country Gentleman*, carried botanical articles related to their specialties. *Godey's Lady's Book* and the *Youth's Companion*, two of the most widely circulated journals of the century, regularly offered women and children botanical information. Journals like the *North American Review* and the *Southern Literary Messenger* frequently included detailed reviews of American and European botanical works.

While the botanical information distributed through popular journals lacked the sophistication of the *American Journal of Science*, it reached far more readers. Often these journals distilled ponderous tomes into more accessible essays. Few amateurs would have cared to read the entire trio of lengthy works reviewed in the January 1834 issue of the *North American Review*, but the thirty-page essay set out nicely a subject of great interest to amateurs and professionals alike: the relative merits of natural and artificial systems of classification.[20] This timely article informed readers of current advances and their potential importance. It argued that the artificial system of classification developed by the Swedish botanist Carl Linnaeus and the natural systems being developed by French botanists both had useful functions. The Linnaean system, which required only counting a few gross floral parts, provided a ready key to the plant world for amateurs; the newer natural system served the professional's needs by accommodating larger numbers of species with greater sophistication. While some die-hard adherents to the Linnaean system worried that adoption of the natural system threatened amateurs, because it made botany too difficult for all but specialists, this article stressed the appropriateness of each for specific uses.[21] This moderate view that recognized a role for both systems was far more common in the literature of antebellum popular botany than was the hue and cry about the natural system threatening botanizing.[22] Journals gave the issue a forum, and kept their readers informed about other changes on the cutting edge of botany.

Mid-Century

By the mid-1840s, the atmosphere of collegiality was changing as the first generation of professionals sought to distinguish themselves from amateurs. Training and employment increasingly joined expertise as characteristics indicating status in the botanical community. While the new professionals had no desire to dispense entirely with amateurs, the perceived "proper" role of the amateur changed from colleague to helper. The flow of information was affected by the increasing assumption by professionals and amateurs alike that professionals should set the agenda not just for themselves, but for amateurs as well, by assuming positions of leadership in the scientific community and using them to shape the style of botany.[23] During the middle decades of the century, professionals took charge of societies and advised amateurs through journals and clearinghouses. Two institutional developments exemplify this change.

Institutions

Characteristic of the new hierarchy was the creation of the American Association for the Advancement of Science (AAAS) in 1848, the first American scientific society to be national in deed as well as in aim. Its membership was a heterogeneous group of professionals, amateurs, and the interested public. Amateurs often joined when the annual national meeting was nearby, and then dropped out as their interest waned.[24] From the beginning, however, the leadership of the society was far less heterogeneous: virtually all of the leaders held jobs that involved science, predominantly teaching or working for the government. The leaders struggled to impose professional standards and ideals of scholarship and documentation upon the papers given at meetings and published in the proceedings. Whether this was an attempt to upgrade the quality of scientific work or an attempt to gain professional autonomy, the result was that amateurs were unable to participate fully in the association.

Nowhere was the professional view of the proper role of amateurs more clearly expressed than in A. D. Bache's presidential address delivered to the AAAS in 1851:

> The world is made up of ordinary men, and it is part of common sense not to despise their doings. The specimens collected or the observations, made by the humblest geologist who ever wielded a hammer, or the meekest astronomer who ever noted a transit, serve as part of the foundation of the superb structure raised by Van Bach, or by Leverrier. If the zeal of *second-rate* men is warmed into activity and directed in the development by such influence, the general level of science is raised by slow deposits, which may on occasion make mountains by upheaval.[25]

Of more direct importance to botanizers than the AAAS was the creation of the Smithsonian Institution in 1846.[26] The Smithsonian Institution, with Joseph Henry, a pioneer in electricity and magnetism, at its helm, rapidly developed a reputation as a major source of support for amateurs in all areas of science throughout the last half of the nineteenth century. The Smithsonian provided supplies, expense money, and advice to collectors throughout the country. So many amateurs wrote for instructions on collecting that the Smithsonian printed them for distribution. Among the most valuable services the Institution provided were purchasing books for those in remote areas, introducing correspondents to others with similar interests, and serving as a mail drop for traveling naturalists.[27]

Clearinghouses

The emergence of American societies and journals supplemented rather than replaced the less formal mechanisms of communication of the colonial era. The natural history circle was a casualty of the Revolution. In its wake, there arose American botanical clearinghouses.[28] The earliest of these was conducted by John Torrey (1798–1873), who, while never able to make a living on botany alone, devoted much of his time and energy to the study of plants. Being unable to afford the expense or time of extensive fieldwork, he often sought specimens and observations from those who could travel or do fieldwork in other regions; as a result, he developed an extensive circle of correspondents who exchanged specimens for expertise, or for specimens in trade. Because Torrey's work was dependent upon amateurs, he seldom begrudged a request for help and he regarded even the most nascent of amateurs as colleagues. Among the protégés whose development as botanists he helped shape were a few individuals who became professionals, like Asa Gray. Many more, like Hardy B. Croom, remained amateurs. Torrey offered one and all the advice and wisdom of his experience.[29]

Writing to a more experienced botanist was a natural step for an amateur in search of proficiency. Sooner or later, even the most dedicated amateurs were likely to find specimens they could not identify and to then want someone more knowledgeable to have a look. Often the expert of first resort was a regional authority—for example, Increase Allen Lapham of Milwaukee, a self-educated land speculator and developer and a prominent figure in the upper-Midwest of his day (1835–75), whose interest in natural history, especially botany, was well known. Amateurs in a variety of disciplines from throughout the region wrote him for advice and assistance, sending a specimen with the query, "Can you identify this?" or asking how to begin the study of botany.[30] Lapham's response took one of two forms. If, on the one hand, the correspondent was interested in whatever discipline currently occupied Lapham, he would take that individual under his own wing, trading specimens for identifications, suggesting or providing identification guides, and even visiting to lend moral support. If, on the other hand, Lapham was not personally interested, he recommended the inquirer to whichever of his many Eastern correspondents would be most intrigued, thereby incorporating the individual into a national network.[31]

Lapham himself corresponded regularly with two of the major national botanical clearinghouses, Asa Gray at Harvard and S. F. Baird at the Smithsonian. The Smithsonian was happy to receive Lapham's contributions, and reciprocated with publications, but the Institution's general lack of interest and expertise in botany was evident in its dealings with Lapham as well as with others. Assistant Secretary Spencer Baird wrote to John Torrey on several occasions to solicit help with specimens, including a query about the identity of a tree from his own back yard, which suggests that the Smithsonian staff was deficient in botany. This weakness in botany was reflected in an 1859 letter from Baird to Torrey that read:

Dear Doctor,
    I enclose a small package of plants just received from Loramil Peak. Please give me some indication of their character as to interest or rarity as I want to write to the person sending them to get more and to tell him something about them.
                                    Very Truly Yours,
                                    S. F. Baird[32]

Baird was clearly trying to promote botany, but without committing staff time. In some instances his benign neglect of botany went still further. More than one individual, writing for instructions on the collection of plants, received the desired information *and* a request to collect fish, Baird's interest![33] The Smithsonian never begrudged the contributions that amateurs made to its collections, but because it saw those amateurs as aides to science, not as full partners, it felt free to guide them in a way that Torrey did not.

The impact of the new professionalism on amateurs is clearly seen in the Harvard information network, undoubtedly the most important botanical exchange of the mid-nineteenth century. Asa Gray assumed the Fisher Professorship of Natural History at Harvard University in 1842. For the next forty-six years, he used his status as America's most famous and visible botanist to build Harvard's program in botany. His enormous botanical correspondence included men and women, young and old, from California to Maine, who sent specimens in return for help or merely for the purpose of establishing collegial contacts. When an enthusiast first wrote to Gray to ask for help or to offer specimens, Gray, like Lapham, either incorporated the amateur into his own network or recommended another professional who desired specimens from the correspondent's region.

In the case of those he kept in his circle, Gray often had a specific agenda to push or a particular question he wanted answered. When Charles Darwin asked Gray to set the geographic range of species in the latter's *Manual of the Botany of the Northern United States*, Gray recruited Lapham to apply his field knowledge to set the northwestern bounds of a number of species. Gray sought and gained insight into the geographical distribution of species; Lapham saw the issue as one with potential applications to the agricultural development of Wisconsin. Yet although their motives differed greatly, they were compatible, and both men profited from the relationship. This pattern of amateur and professional both profiting, but in very different senses, was repeated again and again throughout mid-century.[34] As the size of the network grew, Gray's colleagues and assistants in Cambridge helped keep up with the correspondence by taking on those amateurs deemed unworthy of the master's time, turning Gray's personal clearinghouse into a Harvard clearinghouse.[35]

One group of botanizers in central California who became part of the Harvard network illustrates especially well the importance of clearinghouses to those in geographically isolated areas. J. G. Lemmon, who took up botanizing and the outdoor life in hope of restoring the health he had lost as a prisoner of war at Andersonville, began to write to Gray from the Sierras in the early 1870s and was soon sending specimens on a regular basis. His letters painted a vivid picture of a highly dedicated amateur living far from the resources and expertise he craved. Gray's periodic donations to cover expenses, his shipments of books and supplies, his expert evaluation of specimens, and his respect and friendship all meant a great deal, as revealed in a quick note to "My Dear Doctor" in 1875: "Home again, lots of letters, lots of work, and courage in plenty! . . . Found many favors awaiting me from you. Money, botanical paper, reports, honors, & c., Thanks, Thanks."[36] Lemmon's "honors" referred to the recent naming of a species for him. Gray later honored Lemmon's wife Sarah Plummer, an enthusiastic and talented botanizer in her own right, in similar fashion, at which point Lemmon reported: "we have just held a grand celebration, mother & I dancing around the room . . . alternately shouting for joy and weeping with gratitude."[37]

The money Gray provided was also crucial in Lemmon's case for he was often in dire financial straights, spending most winters as a caretaker at a resort. Donations from those who wanted specimens allowed Lemmon to purchase minimal supplies and go collecting. It did not take much money to keep a botanizer going: ten dollars that Lemmon passed on to Rebecca Austin enabled her to keep working on a local study of *Darlingtonia*.[38] In the late seventies Lemmon reported delightedly to Gray that his fellow Californians were beginning to seek him out to teach and lecture on botany. Jokingly, he asked Gray: "Do you wish to borrow some money?"[39]

Indicative of Gray's importance to the Californians was the excitement attending a visit he made to California in 1877 with Mrs. Gray and the distinguished British botanist Sir Joseph Hooker. Rebecca Austin, who had by this time begun a correspondence of her own with Gray's Harvard colleague George Davenport, wrote that she had nearly wept when she learned that she had missed Gray and his party. Lemmon, she reported, had met them at the train station at Truckee, a town in Donner Pass, and had spent two days with them.[40] A reference to the visit by Lemmon indicates both the popularity of botany and Gray's fame. In a letter of 6 December 1877 Lemmon reminded Gray of promised photos of himself, Mrs. Gray, and Hooker. Lemmon added that during a recent trip to San Francisco he had found photos of Gray and Hooker for sale "among the celebrities of the U.S. but yours was named 'Sir Joseph' and vice versa."[41] To botanizers like Lemmon, Gray was much more than an expert botanist, he was a cultural hero, who understood and encouraged their passion.

Periodicals

Because Asa Gray wrote for journals as different as the *American Journal of Science* and the *Atlantic Monthly*, his influence permeated both the scientific and the popular presses of the middle decades.[42] The fifties and sixties saw the heyday of popular journals, including old ones like *Godey's Lady's Book* and a host of new ones like the *Atlantic Monthly*. It was during this era that *Godey's* proclaimed botany a subject its readers should find interesting, and hence began including reviews and articles on topics like "Botany as a Study for Young Ladies."[43] *Godey's* was by no means alone or unusual in this coverage. The *Atlantic Monthly* regularly published essays on a range of "nature topics" that indirectly conveyed a wealth of botany; it also published more serious science articles like Gray's series.[44] The *Country Gentleman*, a popular farm journal, was an ardent supporter of the usefulness of botany to agriculture.[45] Children's magazines like the *Youth's Cabinet* included reviews and articles on both botany and the botanical aspects of natural theology.[46]

The botany included in widely distributed popular magazines was tailored to a general audience. While the *American Journal of Science* might publish technical descriptions of new species, the most sophisticated material that *Godey's* could print was a description of floral parts. More common in popular journals like *Godey's* were discussions of the merits of botanizing as a pastime, and "how-to" tips on collection and preservation, all aimed at a relatively inexperienced audience.

Late-Century

The warm cooperative spirit was not to outlive Gray (who died in 1888) by very long. As the century drew to a close two trends changed the character of the American botanical community: professionalization, and a shift from the natural-history focus to a biological focus in botany. As the professional community gained strength, leading botanists sought autonomy—which, of course, precluded any significant interaction with or reliance on amateurs. Simultaneously, the increasing importance of experimental rather than observational work gradually relieved professionals of the need to enlist the help of amateurs.

As in other areas of science, many of the new botanical positions available in the latter half of the nineteenth century were academic, and certainly the botanical leadership was heavily academic.[47] In New York, in Boston, and at the national level, societies were either co-opted by academic scientists or abandoned to amateurs. In the 1890s botanists formed their first professionals-only society, the Botanical Society of America. No longer were professionals and amateurs peacefully and productively coexisting. By century's end, a sizable community of professional botanists was actively struggling for autonomy. Jobs, training, and specialized societies all were in place, organized around a new biologically oriented style of botany that fit the needs and motivations of professionals alone.[48] The free transfer of information and specimens would dwindle because in this new world it served neither group. What had been one community would become two.

3 Botanizing and Self-Improvement

Underlying virtually all of the motivations of amateurs for botanizing, including natural theology and attitudes about work and play, was the belief that nature's "improving hand of cultivation" promoted intellectual development, physical culture, gentility, and practical knowledge.[1] Just as some antebellum Americans promoted improvement through temperance, public education, and health reform, so others—albeit less ardently—championed botany as a means of improving individuals and, as a consequence, society. In an age buzzing with reform, the image of botany as an improving activity served firmly to establish its popularity.

Botanizing outlived many similar antebellum causes, in part because its links with self-improvement survived the social upheaval of the Civil War era; this allowed it to emerge recast in the more modern form of personal advancement that stressed cultural and individual career benefits over societal reform. As those who had hoped to build a more perfect society through individual reform became less vocal, the new promoters of botanizing championed self-improvement as a means of getting ahead in society. Because it could be adapted to meet the changes in how Americans viewed self-improvement, botanizing remained a popular vehicle for personal enhancement throughout the century.

Early Reformism

Antebellum America was a hotbed of moral and spiritual reform movements, so it is hardly surprising that the champions of botanizing of the age cast it in the light of self-improvement. The impetus for self-improvement came from a variety of sources. The strongest threads were religious, stemming from the concept that the moral rebirth of a nation inevitably follows the diffusion of knowledge and culture. Integral to antebellum reform was the religiously founded belief that society can be dramatically improved by its individuals' working for perfection in all aspects of life.[2] The logical end of this thread, seen clearly in transcendentalism, was each individual's obligation to cultivate God-given abilities to his or her fullest potential.

In addition to its religious basis, self-improvement also had a strong economic component. Self-improvement was an inexpensive and readily available means of attaining social and economic advancement by exercising each citizen's "natural right" to knowledge. Additionally, a utilitarian thread in agricultural and medical botany fostered practical knowledge and skills that increased self-sufficiency.[3] Botany was, in the minds of some, "the best corrective of that heartless and demoralizing state of society which consists of personal associations without sympathy; of forms without affection; and of professions without sincerity;—where each have studied and practiced deception, until the vices of each are known, and falsehood can no longer conceal depravity."[4] Just as they promoted temperance and abolition, reformers encouraged others to botanize in order to improve themselves and thereby society through the study of a science that "combines pleasure with improvement."[5]

Enthusiasm for self-improvement through botanizing was great-

"How Plants Grow" (Cover from Asa Gray, *Botany for Young People and Common Schools. Part I: How Plants Grow*, 1858)

est among those concerned about the welfare of the young and the poor. Some reformers began botany circles because they "could not feel at ease anywhere unless the young persons around were cultivating those tastes for simple pleasures which would be to them a source of continual improvement and enjoyment."[6] Another rued the general inattention he saw being given to botany in the schools:

Were children instructed in the economy and habits of plants, and were their minds enlightened in regard to their uses, and the intimate relations which they sustain to the human family, it would open to their minds a source of knowledge which could not fail to exert a most important and salutary influence upon their moral character. Such a knowledge would give them a just idea of their physical and moral relations, and enable them to appreciate the necessity of physical employment, it would be the best field for intellectual improvement and unite the dignity, and refinement of science, with the exercise of labor, and harmonize the physical with their moral natures.[7]

In contrast, the author warned, one not so instructed finds no pleasure in life or work. "His heart is void of gratification; all his aims are selfish, and tend to a state of moral degradation."[8]

These sentiments were by no means unique or even unusual. Throughout the century, articles in popular and scholarly journals, as well as introductions to textbooks—a favorite vehicle for prescription—presented the improving nature of botany in morally laden terms. The *American Journal of Education* in 1829 urged schools to introduce the subject for nine reasons: interest to the youthful mind, creation of a purpose to walk, promotion of exercise, development of observation, instruction in good order and arrangement, utility of the subject, moral improvement of the young, ease of study, and suitability for females. Of these, moral improvement received the most attention and the strongest endorsement: "We do not say that a botanist will never be a narrow minded man: but we believe that the natural tendencies of botany are altogether good."[9] A few years later a children's magazine took up the same theme. Listing four reasons why children ought to study botany, the author stressed the promotion of exercise, the development of the powers of observation and of orderly habits, and the improvement of character: "Botany helps to make the temper mild and agreeable; and has a tendency to refine and improve the mind. It is among the most innocent things in the world, . . . It has a tendency to lead us to think of God."[10] One fictional account cautioned that those mothers who prohibited the pursuit of botany because specimens brought dirt and disorder into the house might "unwittingly be inflicting a most serious evil on their own children, and on society."[11]

One especially strong view of the influence of botanizing on children came from Timothy Dwight, president of Yale University:

The late President Dwight was an eminent champion of the virtue which he practised [sic]. He often directed the attention of his pupils to Sweden, to point out the influence of natural history on the moral character of man. In that country Botany is taught in the schools, and the habituation of her excellent children presents a cheering picture of domestic felicity. Their piety and their patriotism both flow from the same source; for while they examine the productions of their country, they become attached to its soil; and while they contemplate the works of their Maker, they are animated with the glowing spirit of devotion.[12]

Others had so much confidence in the improving nature of botany that they saw it as a panacea for improving the lot of "sons and daughters of Toil," both as a group and individually.[13] Properly taught, one author argued, botany "cannot fail to add immensely to the material wealth, the intellectual and aesthetic culture, and thus to the happiness and general welfare of the community."[14] Another wrote that the individual laborer would find it a relief "to converse with and receive instruction from those things which are connected with his daily employment; instruction too, so pure, so refined, and so unalloyed by human passion, that it cannot fail to improve his heart and make him a better man"— indeed, the laborer who resisted such instruction could not "expect to maintain a position in the progressive improvement of the age equal to that of those who are engaged in other pursuits; and instead of occupying, as he should, the highest position in society, he must take his station in the lowest."[15] Yet another enthusiast of the civilizing powers of botany, while perhaps more optimistic than most, expressed the fondest dreams of a generation of botanical writers: "The reader of only ordinary education and intellectual powers may readily comprehend the principles inculcated in this book, and see their applicability as guiding rules for the judicious and happy management of each day's duties. . . . may you be induced . . . to look upon the tree with new interest; and obtaining from its noble form a clear and truthful view of your own position and duty in life, become by the perusal of this volume, a wiser man and a better citizen."[16]

The promoters of botanizing fit the science into the mold of reform by stressing its improving attributes, especially its genteel nature, its practical applications, and its many links with natural theology. Gentility encouraged the more refined, more respectable characteristics embodied in moral regeneration. Practical botanical knowledge would make better farmers, better purveyors of home medicine, and generally better citizens. Natural theology would lead to greater piety. Gentility, utility, and piety were familiar themes in mid-century America that appealed strongly to the moral leaders of the emerging middle class. By fitting botanizing into this common framework, its champions aligned it closely with the prevailing reform spirit.

Of these mid-century claims to self-improvement, only natural theology pervaded the botanical literature more thoroughly than gentility. The cult of gentility, a set of social codes rigidly defining behavior, appealed particularly to the upper half of society from which most botanizers probably hailed. Spread by middle-class moralists who sought to create a world of piety and refinement, gentility was a compelling call to botanize. For those already within genteel culture, the components of gentility—refinement, respectability, politeness, and elegance—figured heavily in choices about what one should and should not do. Those hoping to become more genteel chose activities on the basis of cultivating gentility. In either case, botany proved an attractive activity.[17]

The qualities of botanizing that made it genteel were many and varied. In the most general terms, botany's "improving hand of cultivation . . . imparts a taste for the beautiful in nature, fills the soul with ravishing emotions, and opens before us the enchanted avenues of a world full of wonders."[18] More specifically, botanizing was conducive to other genteel activities, including the arts of flower painting and the construction of wax flowers, both popular nineteenth-century hobbies, especially among women.[19] It enhanced "poetic sensibility," and was useful in both constructing and interpreting literary symbolism.[20] Botanizing was helpful in writing accurate travel memoirs. It was associated with the "language of flowers," which attached symbolism to particular flowers sent from one person to another.[21] Flowers—live, fresh-cut, and dried—were also widely used for decorating.

Unlike zoology, botany did not require chasing or killing animals. As one author who advocated botany over zoology put it: "Besides, it is impossible to study the history of sentient beings without some degree of cruelty, which, however justifiable, nay laudable, when exercised by philosophers in scientific investigations, is undoubtedly to be counted among the repulsive circumstances in the study of a great part of natural history."[22] This complaint ruled out even ornithology, which stressed collecting specimens until binoculars became popular and affordable after the Civil War, and entomology, because it required "the bottling of poor beetles in spirituous solutions, the pinning of innocent moths and gay butterflies, and other cruel operations."[23] Unlike geology, botany did not involve dragging around heavy rocks or enduring strenuous fieldwork. One author summed up the appeal of botany especially eloquently: "In botany all is elegance and delight. No painful, disgusting, unhealthy experiments or inquiries are to be made. Its pleasures spring up under our feet, and, as we pursue them, reward us with health and serene satisfaction. None but the most foolish or depraved could derive anything from it, but what is beautiful or pollute its lovely scenery with unamiable or unhallowed images."[24] There was, in short, "something of grossness, of ungrateful toil, involved in the most moderate attention to all other branches of natural history, which deserves to be had in consideration in estimating their relative claims to general popularity."[25] Botany, in contrast, was "among the most innocent things in the world."[26]

In addition to the prevalent theme of gentility, two newer aspects of self-improvement, mental discipline and physical culture, began to emerge during the antebellum period. While the rhetoric behind gentility most often stressed antebellum notions of societal reform, the rhetoric of mental and physical culture stressed personal benefit, a motivation that would blossom more fully in post–Civil War America. During the antebellum period, however, both mental and physical culture were also allied closely with schooling, a mainstay of societal reform. Educators introduced botany and other branches of natural history into the curriculum in part because of their gentility, but primarily because they provided good mental and physical exercise, important considerations in educators' goal of producing better citizens.

Educators were especially interested in botany as a means of improving mental function through developing the powers of observation, memory, and reasoning. Traditionally, schools used the classics and mathematics to develop "mental discipline," but increasingly in the nineteenth century science became a valued addition to curricula, because of its effects on intellectual development.[27] Botany, advocates argued, had the same beneficial effects as the classics or mathematics on developing reason and system, but was far more pleasant to study:

> The love of flowers seems inherent in childhood; their bright colors and beautiful variety attract the eye at a tender age, and if, in exercising the memory and training the intellect, botany can gain a stand-point upon the platform now universally conceded to the Latin grammar and mathematics, much will be accomplished; not that we would depreciate the advantages arising from the study of the dead languages, particularly the Latin, which is so necessary in the science under consideration; but how many there are who have a decided dislike for those studies, and whose leisure hours are a waste and weariness to themselves and their friends, yet who might have become proficient in botany, had their attention been early attracted and their minds trained by the study of this most charming and instructive of the natural sciences![28]

By the 1830s, educators promoted natural history as the study "best suited to the exercise of several powers of the mind" and as "a good mental exercise for youth."[29] Because of the systematic focus of botany "in which you are obliged to sort, and arrange, and classify things," antebellum enthusiasts told their audiences that botany would encourage order and system throughout their lives: "Some boys and girls never have any particular place for everything, and so when they wish to find a needle, a pencil, a book, a pair of scissors, or a penknife, they do not know where to find it; and a great deal of time is often wasted in looking for it. This is a bad habit, but it is one which the study of Botany will help to correct."[30] In an age that equated orderliness with good work habits and morality, the argument that botany trained the mind and promoted orderly thinking led to the belief that "Botany deserves our highest regard as the source of mental improvement."[31] Botany was being taught not for its content so much as for the way it "accustoms the mind to systematic arrangement, definite rules of classifications, and strict attention to the import of terms."[32]

In addition to encouraging the development of orderly thinking, botany developed valuable observational skills. Nineteenth-century American texts for children admonished their readers to become more observant: "Many boys and girls go through the world, almost without seeing it. Now he who has eyes and does not use them, in such a beautiful world as this, is very much to be pitied. But the study of Botany will learn him to keep his eyes open. The habit of noticing things around you is called observation; and a very important habit it is, too."[33] By going "directly to nature" to observe, rather than studying plants from a book, botanizers not only gained botanical understanding but also became more interested in the world.[34] Thus, botany became a tool for intellectual enhancement, with the ultimate goal of creating orderly minded, curious individuals.

Similarly, antebellum educators and reformers hailed the gentle exercise incident to botanizing as a cure for the problem of an increasingly sedentary society. Nineteenth-century Americans regarded walking as healthful exercise, because it was not overly strenuous. Since "the love of plants calls us into the fields," searching for botanical specimens was an inducement to exercise.[35] The incentive of collecting would draw the botanizer from the study or parlor to the woods and meadows, with great benefit to both mind and body. Enthusiasts claimed that "Botany affords precisely the inducement to leave the close air of the school room, the study, or counting room, and to wander in search of the fairest objects of inanimate creation."[36] Some botanizers found the lure of fresh air nearly addictive: "When I say that the study of Botany is healthy, I mean that it will lead you into healthy practices. You will be abroad every opportunity you can get, roving among the trees, in the fields, and on the mountains. You will be glad to do so to breathe the fragrant and pure air. And it is healthy."[37]

Postbellum Personal Advancement

The social upheaval of the Civil War tolled a death knell for the antebellum style of reform.[38] In place of societal reform through individual improvement, there arose a new faith in personal advancement based on hard work and self-discipline. In this new spirit, botany became a tool for improving the quality of one's own life rather than a means of reforming society. One particularly clear statement of this new attitude came from "an ignorant mechanic, deaf and without even a common school education," who wrote a botanical autobiography in the hope of "encouraging others in greater diligence, patience and perseverance."[39] Botanizing had opened new worlds to him, giving him great pleasure and a new circle of acquaintances. While this worker clearly felt that botanizing had improved his life, the improvement came not in the old form of being a better citizen, but rather in the newer form of personal advancement, of rising above his station of "ignorant mechanic."

The appeal of botany as a curricular item so evident before the Civil War not only survived the decline of antebellum reform but reemerged with new strength. Post–Civil War educators compared botany favorably to the old curricular standards "in the training it gives to the powers of observation and critical judgement," while noting that it appealed to students more than the classics.[40] This was not an entirely new line of argument. Thomas Sherwin, the principal of Boston English High School, had echoed a generation of educators in the fifties when he wrote that, "as a mental discipline, the study of science may boldly challenge comparison with that of the classics"; he went on, however, to say that the real benefit of substituting science for Latin or Greek was that since the boys were more interested they were transformed from discipline problems to having "no time for mischief."[41] Post–Civil War educators, in contrast, were less apt to describe the salubrious effect upon the classroom or society than to stress the benefits to the individual student.

Like their predecessors, advocates of botanizing after the Civil War saw the systematic approach to identification as encouraging "the closest reasoning, the most thoughtful weighing of evidence, [and] the acutest application of the logic of facts."[42] What late-nineteenth-century educators stressed most, however, was the belief that the logic learned would lead to intellectual growth. While antebellum children had read that the orderly thinking obtained through the study of botany would help develop the good habit of having a place for each possession, late-century readers learned of the effects of botanizing on the powers of reasoning through the often-cited example of John Stuart Mill's practice of botanizing "with a view to its important mental advantages."[43] This line of thought reached its high point in the popular works of Eliza A. Youmans, whose introductory botany text bore the subtitle "Designed to Cultivate the Observing Powers of Children."[44] Youmans argued that "the habit of systematic arrangement in which the study of botanical classification affords so admirable a training, is equally valuable in methodizing all the results of thought."[45]

The support for botanizing as a form of exercise also changed to fit the new emphasis on discipline as a tool to enhance personal well-being and advancement. One physician favorably compared the ability of botanizing to harden the muscles and strengthen the frame with time devoted to "the bat, the racket, or the bicycle," diversions that were all the rage in the Gilded Age.[46] A prominent educator and scientist stressed exercise as a benefit of botanizing for teachers: "Many . . . teachers, especially females, are breaking down from time to time, for want of air, exercise, and sunshine. The looking for objects for their lessons in natural history would . . . oblige them to take long walks."[47] Exercise was less the duty it had been in the antebellum era, and more a requisite for personal well-being.

The use of botany to promote well-being shows clearly in the growing reputation of botanizing as an ideal activity for invalids in need of "mental distractions" and gentle physical exercise. While there are certainly accounts of antebellum invalids botanizing, late-century enthusiasts recommended it because it encouraged moderate exercise in the fresh air while it amused and entertained.[48] Botany necessitated no cumbersome equipment and was relatively easy to study, which made it ideal for the recovering patient "supposed to be strong enough to do a reasonable amount of walking and some solid thinking."[49] One advocate spoke from experience: "the most dismal moments I have ever spent were while laboring conscientiously with dumb-bells and Indian clubs in the name of exercise. Physical exercise for its own sake, is intense and profitless, and often, I believe, pernicious labor. Give yourself a motive for exertion, and it then becomes exhilarating. The study of plants supplies just such a motive as invalids need."[50]

Late-century Americans who could afford it convalesced at rest homes, water cures, and spas, and enthusiasts therefore urged such health resorts to promote botanizing among their patients. One proponent credited botanizing with awakening the invalid's motivation to exercise and arousing enough enthusiasm to "put more life into a sick body than all the drugs in the dispensary."[51] At least one such institution, the Poland Springs Water Cure, hired a botanizer to serve as house botanist for the "amusement of the half invalid."[52] Botanizing was far less dismal than toiling at calisthenics and provided a "wholesome diversion from the imbecile fancy work, and still more imbecile gossip" that occupied the lives of most "resort-bound invalids."[53]

The portrayal of botanizing in fiction as therapy for invalids provides powerful evidence of the depth of its popularity. Elizabeth Stuart Phelps's Doctor Zay brings her patient, Waldo York, a sensitive plant and offers to lend him Darwin's book on the power of movement in plants in her only attempt to entertain him.[54] Louisa May Alcott created botanizing invalids in both *Eight Cousins* and *Under the Lilacs*. In *Under the Lilacs*, the sickly boy Thorny enlists the collecting services of the impulsive lad Ben; in the course of his convalescence, Thorny gradually does more and more of his own collecting as he regains his health.[55] Alcott's prescription of botanizing for invalids mirrored real life, as shown by the number of individuals who took up botanizing as a counter to ill-health.[56]

Invalids were not the only fictional botanizers of the late nineteenth century. Alcott's Ben begins collecting as a favor for his invalided neighbor. Eventually he becomes interested in botanizing, with a little encouragement from coming to understand the value of being able to identify poison ivy. In the process, he loses some of his mental laziness, becoming more inquisitive.[57] Similarly, another fictional child, young Harry, set by his aunt to making leaf prints to while away a hot summer afternoon, becomes interested in the leaves themselves and hence in botany; soon, his sister complains that he "pores over his chemistry and botany to find the why and wherefore of everything" and "has all his leaves arranged in classes, always finds some new specimen, torments us with terrific chemical, botanical, and dictionary words."[58] Other children learned to overcome their shyness by managing to talk about botany with adults, or were understood to be the "right sort" because they botanized so tirelessly.[59] The surprising frequency with which botanizing occurs in late-nineteenth-century fiction and the uniformly favorable light in which it is portrayed speak clearly about its status.[60]

The nature of botanizing's reputation changed, however, as the century came to a close. The muting of the appeal of gentility and utility was but part of a broader decline in the interest in botany as a means of self-improvement. Those, like Youmans, who argued that botany developed the mind, shared the press with others who, despite advocating botany as a school subject, pointed to the important role of the classics and mathematics.[61] And new attractions like bicycling and baseball made the exercise of collecting seem tame and dull by comparison. Botanizers could still argue at the turn of the century that botany was a subject that benefited them, but not with the convincing certainty or the single-mindedness of a half-century before. "The improving hand" of botanizing, while still an important force, no longer drove the interest of the amateur community quite so clearly.

4 Children, Education, and Amateur Botany

During the nineteenth century, schools of every kind—common schools, academies, seminaries, the new high schools, colleges, and medical schools—served as important institutional homes for botany.[1] Schools not only provided training and jobs for those who wished to pursue botany vocationally, but also offered botany to students who had no scientific aspirations. Indeed, schools and textbooks were the media through which many, if not most, botanizers first encountered the science. Throughout the century, educators and scientists debated at what level and by what methods to teach botany—and most basically, *why* botany should be taught. These debates about the content, methodology, and philosophy of school and textbook botany both reflected and influenced changes in botanical expertise, pedagogy, and the perceived value of botany.

From its introduction into curricula in the 1820s and 1830s, botany remained a standard subject throughout the century. The goals behind its inclusion in curricula and the methods used to teach it, however, varied considerably from school to school, from teacher to teacher, and especially from decade to decade. Nonetheless, patterns do emerge. Debates between scientists and educators over the inclusion of botany in secondary-school curricula provide perhaps the clearest example of the differing goals of educators and scientists. Those whose primary mission was education used botany to teach a variety of skills and values. In contrast, those who actively pursued science as a profession sought to teach botany on its scientific merits. The tension between these very different and increasingly incompatible goals shaped the form of botany in secondary education.

Botanical education went through three distinct phases between its introduction into the curriculum and the start of the twentieth century. During the first era, which ran from the 1820s to the 1840s, educators saw the worth of botany in the skills and values its study enhanced—observation, love of the Creator, and analytical thinking—rather than in scientific knowledge itself. The educators of the first era, led by Almira Hart Lincoln Phelps and Amos Eaton, were more interested in pedagogy and morality than in science. They taught students to identify plants using the artificial system of classification developed by Linnaeus, defending it as an educational tool despite its well-known scientific flaws. The botanical educators of the second era (running from the 1840s to the 1880s), epitomized by botanist Asa Gray and educator Alphonso Wood, did not deny the training value of botany, but placed a far greater value on the intrinsic merit of botany as a body of knowledge. Striving for more scientific accuracy, mid-century educators and scientists introduced natural systems of classification. During the third era, which had its roots in the 1880s but did not reach its apex until well into the twentieth century, botany broke into two parallel crusades: the educator-backed "Nature-Study" movement, which stressed nature appreciation for grade schoolers, and the scientists' "New Botany," which stressed scientific thought and method in secondary and postsecondary education. The third era focused less on taxonomy and more on the physiology, morphology, and development of plants and plant communities.

"Uncle Philip Talking to the Boys" (Frontispiece from Francis Lister Hawks, *The American Forest: Or Uncle Philip's Conversations with the Children about the Trees of America*, 1834)

Changing Curricula

In the early nineteenth century, formal opportunities to study botany were limited. Medical studies often included it because of the heavy use of plant-based drugs, and the handful of American colleges occasionally offered a course to their small pool of students. As can be seen in table 1, the presence of botany in an academy's or seminary's curriculum was unusual, but not unheard-of. During the late twenties and the thirties, however, social factors as diverse as a reawakening of evangelical religion and the beginning of urbanization coalesced to create a greater interest in both privately and publicly sponsored schooling.[2] Secondary education flourished and the number of academies and seminaries expanded exponentially, touching a relatively large number of students. It was in the secondary school that botany found its educational home throughout the nineteenth century, becoming a standard offering there at least a decade earlier than in the college.[3] The number of secondary schools offering botany provides one of the clearest, most dramatic indicators of the popularity of botany (see table 2).

Just as the number of academies and seminaries grew, the range of subjects offered also expanded.[4] New York seminaries and academies, for example, reported just fourteen subjects taught when record keeping began in 1787, and only nine new subjects had been added by 1824. Between 1825 and 1830, however, fifty-eight new subjects joined the list. The only sciences that New York schools taught prior to 1826 were natural philosophy (now called physics), astronomy, and chemistry. Geology, mineralogy, mechanics, natural history, and botany appeared in a few schools in 1830.[5] While New York provides a good example because of the unusually good records kept there, it would be a mistake to generalize too firmly to the rest of the nation. We do know that New York was not alone in beginning to offer botany, as existing figures from other states, shown in table 2, indicate. In Massachusetts, for example, many schools began to offer botany, and by 1857 the state required that it be taught in the bigger towns and cities.[6] New York's good records of curricular offerings, however, make it possible to pinpoint the 1830s as the decade during which botany entered the secondary-school curriculum and became a standard subject, as shown in table 3.

Table 1
Academies and Seminaries Offering Botany prior to 1830

School Sample	Percentage Offering Botany
Pennsylvania, boys or coed (1750–1829)	0
Pennsylvania, girls (1750–1829)	14
New York (1800–30)	0
Female seminaries, national (1749–1829)	24

Sources: James Mulhern, *A History of Secondary Education in Pennsylvania* (1933; rpt. New York: Arno Press, 1969), 328–29, 428–29; George Frederick Miller, *The Academy System of the State of New York* (Albany: J. B. Lyon, 1922), 108–9; Thomas Woody, *A History of Women's Education in the United States* (New York: Science Press, 1929), 2:418.

Notes: Despite Miller's claim, I know of at least two schools in New York that *did* offer botany earlier than 1830—Rensselaer and Troy Female Seminary—but it seems safe to conclude that these were the exceptions rather than the rule.

There is no guarantee that a school reporting the teaching of botany in fact offered the subject every year, as both Mulhern's and Woody's samples looked at random years. Also, in any of the schools, botany may have been offered but not required. Nonetheless, the statistics offer a yardstick for comparing different eras, genders, or branches of science.

Table 2
Academies and Seminaries Offering Botany after 1830

School Sample	Percentage Offering Botany
Pennsylvania, boys or coed (1830–89)	33
Pennsylvania, girls (1830–89)	77
New York (1830–90)	65
Female seminaries, national (1830–71)	82

Sources: Mulhern, *Pennsylvania*, 328–29, 428–29; Miller, *New York*, 108–17; Woody, *Women's Education*, 418.

Table 3
New York Secondary-School Offerings of Botany

Year	Schools Reporting	Schools Offering Botany	Percentage Offering Botany
1830	58	8	14
1835	66	30	45
1840	127	94	74
1845	153	113	74
1850	166	116	70
1855	164	134	82
1860	192	138	73
1865	202	142	70
1870	182	123	68
1875	216	131	61
1880	237	155	61
1885	261	144	56
1890	335	206	61
1895	504	288	57
1900	705	373	53

Source: Miller, *New York*, 108–9.

The marked rise in the percentage of academies and seminaries teaching botany between 1830 and 1840 was not unique: other disciplines, especially astronomy and chemistry, also grew in popularity during the same period (see table 4). The sciences, enjoying the reputation of being practical and conducive to piety, benefited generally from the antebellum awakening of interest in learning.[7] Chemistry, natural philosophy, and astronomy were the most commonly offered secondary-school sciences during the period from 1830 to 1870. The rise in the popularity of botany, however, was far greater than that of any other area of science. As can be seen in tables 5 and 6, botany was the most popular division of what was then called natural history (the observational rather than experimental study of plants, animals, and other objects

Table 4

*Increase in Rates of Science Teaching in New York
Secondary Schools*

Discipline	Percentage of Schools Teaching	
	In 1830	In 1840
Botany	14	74
Astronomy	40	87
Chemistry	50	93
Geology	2	28
Zoology	0	0
Natural History	16	23

Source: Miller, *New York*, 108–9.

Note: Some of the courses listed as natural history *may* have been largely zoological as the two terms were sometimes used interchangeably.

of nature)—far more apt to be taught than zoology and rarely matched by geology, especially at female seminaries.

Although it is obvious that botany was commonly taught, it is less clear how many students studied it and for how long. Many schools offered half-year or full-year courses in botany, but some offered only a third of a year, while others offered more than a year. Similarly, some schools required the study of botany, while others offered it as an elective or made it available for an extra fee.[8] Additionally, the length of the school year varied from institution to institution.

Eaton and Lincoln Phelps

While the dramatic introduction of botany into the secondary-school curriculum in the antebellum years was in part due to a general upswing in the popularity of science during this period, the ministrations of influential educators also contributed to the favored status of botany. It is easy to overestimate the influence of individuals; nevertheless, the history of the place of botany in

Table 5

Percentage of Secondary Schools Offering Various Sciences

School Sample	Botany	Natural Philosophy	Astronomy	Chemistry	Geology
Female, national (1830–71)	82	90	85	90	60
Pennsylvania, girls (1830–89)	77	88	67	72	33
Pennsylvania, boys and coed (1830–89)	33	54	47	56	28
Pennsylvania, public (1836–75)	25	87	54	65	33

Sources: Woody, *Women's Education*, 418; Mulhern, *Pennsylvania*, 328–29, 428–29, 542–43.

nineteenth-century American secondary education serves as an illustration of the impact individuals can have. From the introduction of botany into the secondary-school curriculum until the 1880s, a small core of educators dominated the field through the production of influential textbooks, and through the training of students who spread their mentors' word. These individuals, beginning with Almira Hart Lincoln Phelps and Amos Eaton, were pivotal in popularizing botany and shaping the course of amateur practice.

The collaborative work of Eaton and Lincoln Phelps began in Troy, New York, in the 1820s. Reflecting the nation's new interest in self-improvement, Troy was home to a lyceum, a museum of history and natural history, a theater, five new district schools, and a host of other cultural and educational institutions.[9] Young men and women throughout the Northeast flocked to Troy to study at two of the country's most renowned schools, Rensselaer Institute and Troy Female Seminary. Among the instructors were Eaton of Rensselaer and Lincoln Phelps of Troy Female. The texts written by these two dominated the market from the late twenties until the mid-fifties, being read by hundreds of thousands of Americans. Their students at Troy, at Lincoln Phelps's Patapsco Female In-

Table 6

Percentage of New York Secondary Schools Offering Various Sciences

Year	Botany	Natural Philosophy	Astronomy	Chemistry	Geology
1830	14	77	40	50	2
1835	45	100	83	95	5
1840	74	100	87	93	28
1845	74	96	84	90	35
1850	70	95	91	87	31
1855	82	96	93	79	38
1860	73	92	77	79	41
1865	70	95	67	70	42
1870	68	92	65	67	34

Source: Miller, *New York*, 108–9.

stitute in Maryland, and at the countless cities and towns where Eaton lectured, in turn went on to teach, spreading the gospel of botanizing. Through their innovative teaching methods, botany became a popular curricular item, especially for females.[10]

Lincoln Phelps arrived in Troy in 1824 to teach at the female seminary run by her sister, Emma Willard. In Troy, she found Amos Eaton teaching botany and other areas of science not only at Rensselaer but at Troy Female as well. She soon began attending Eaton's lectures on botany, and immediately embraced both the subject and Eaton's method of teaching it. Under the direction of Stephen Van Rensselaer, the founder and patron of Rensselaer Institute, Eaton employed the innovative methods of Johann Heinrich Pestalozzi, a Swiss educator—which included supplementing the teacher's presentations with lectures and demonstrations given by students. Ordinarily, education in the twenties and thirties was a matter of rote memorization and recitation. A sample lesson plan published in the *American Journal of Education*, for example, suggested that the class should first recite from their botany text and then discuss the new vocabulary involved, preferably while referring to a specimen of a familiar plant as follows:

Teacher. "How many *stamens* do you find in this flower?"
Pupil. "Five."
T. "It is therefore of what class?"
P. "Of the fifth class, called *Pentadaria*; and of the first order, because it has but one pistil."[11]

The approach at Troy stood in strong contrast. In the case of botany, students at Troy Female and Rensselaer began by learning to identify plants from specimens, rather than memorizing sections of textbooks.[12]

With Eaton's encouragement, Lincoln Phelps soon took over science teaching at Troy Female. Consistent with Pestalozzian pedagogical philosophy, Lincoln Phelps and Eaton both saw botany as a vehicle for teaching a great number of other things that they valued at least as much as botany itself. Botany trained and disciplined the mind to deal with information in a logical way, to observe critically, to analyze—in short, to think. Collecting specimens provided exercise, ranging from moderate to strenuous as suited the botanizer, and helped protect the overtaxed body of the scholar from the mental strain and physical confinement incident to the classroom or library. Further, the study of botany was certain to inspire the student to look from nature to nature's Creator with new love and respect.[13]

Lincoln Phelps and Eaton certainly were not unique in their belief that botany promoted mental discipline. Educators had already begun to use natural sciences to supplement or supplant classical studies, which had traditionally played the role of training the mind. The manifestation of this best known to historians of science is the "Yale Report of 1828," which encouraged the inclusion of natural sciences in the Yale curriculum.[14] Educators interested in encouraging mental discipline especially favored botany because of its taxonomic nature. Not only did botanizing promote "habits of attentive and accurate observation," but it also led to "habits of orderly arrangement," which have "a favorable effect on the mind of the learner."[15]

In 1829, Lincoln Phelps's frustration over the lack of a suitable text led her to publish her botany lectures;[16] her *Familiar Lectures on Botany* was followed in 1833 by a simplified version, *Botany for Beginners*. Together with Eaton's *Manual of Botany for the Northern States* (first published in 1817), which, although less useful as

Table 7

Percentage of New York Secondary Schools Using Major Texts

Year	Lincoln Phelps	Eaton	Wood	Gray
1835	54	20	—	—
1840	65	20	—	—
1845	78	13	—	1
1850	60	4	30	2
1855	(information not available)			
1860	11	—	63	25
1865	3	—	58	39
1870	—	—	45	54
1875	—	—	33	57
1880	—	—	32	65
1885	—	—	14	79

Sources: "Annual Reports of the Regents of the University of the State of New York" (held at the State Historical Society of Wisconsin).

a classroom tool, went through eight editions between 1817 and 1840,[17] Lincoln Phelps's works captured the textbook market for years, as seen in table 7. *Botany for Beginners* went through at least thirteen editions and twenty-six printings, selling (by its author's estimation) 270,000 copies by 1867.[18] The real success, however, was Lincoln Phelps's *Familiar Lectures*, which had at least thirty-nine printings in some twenty-nine editions and sold over 375,000 copies.[19] While Eaton's *Manual* was popular and was certainly influential, its impact pales in comparison to that of *Familiar Lectures* or *Botany for Beginners*. Widely praised in journals as different as *Godey's Lady's Book* and the *National Quarterly Review*, Lincoln Phelps's texts enjoyed extensive use in secondary schools and even colleges.[20] More than a few individuals who became prominent in scientific circles reminisced that their introduction to botany was through one of these two authors,[21] and far more who may never have seen another botany text retained a love of the subject for life.[22]

On the surface, the writings of Lincoln Phelps and Eaton seem to have little in common. Eaton's *Manual* was wholly a field guide to flora, designed to be used by those with some knowledge of plant physiology and anatomy. In contrast, *Familiar Lectures* (and the elementary version, *Botany for Beginners*) included both rudimentary anatomy and a field guide for the near novice. Lincoln Phelps's need for a new and different sort of textbook arose from her conviction that students should study nature directly, rather than relying solely on the authority of others. Although Eaton's *Manual* provided a means for students with some background to go directly to nature, Lincoln Phelps's beginning students needed a more elementary foundation, and her books thus assumed less background. Beneath these differences in format, however, the works of the two authors shared striking similarities. Their common use of the artificial Linnaean system of classification, and the underlying philosophy of their teaching methods, formed a united and influential force.

Throughout their writings, Lincoln Phelps and Eaton both stressed that the process of identifying a plant exercised the powers of observation and developed the skills necessary to distinguish the important from the trivial. In fitting a plant into a system of classification, the student employed logic. Additionally, in recognizing another member of the same species, the student called upon memory and the ability to generalize from an individual to its species.[23]

The pedagogical aim of mental enhancement dictated that the system of classification be a simple one like the artificial system of the Swedish botanist Carl Linnaeus, which enjoyed widespread use in early-nineteenth-century America. The Linnaean system, like other "artificial" systems, is based upon arbitrary segments of the plant (in this case, the floral parts) rather than on the entire plant. "Natural" systems, on the other hand, attempt to group together plants with broad similarities; because they are based upon the entire plant and thus take many more factors into account, these systems are a great deal more complex.[24] In the view of Eaton and Lincoln Phelps, the detail and completeness of natural systems made them better tools for the "learned naturalist" aspiring to further science with new discoveries; beginning students, however, neither needed nor would benefit from anything but the simplest system.[25]

The Linnaean system did have pedagogical merit. Using either Lincoln Phelps's or Eaton's texts, and with a minimum of experience, students could easily identify many common Eastern plants and learn a certain form of systematic thinking in the process. By counting floral parts and in some cases noting other gross morphological features, an average student could determine a plant's Linnaean class and order with relative ease and certainty. The generic and specific identification was then simply a matter of comparing the specimen with descriptions of the possible choices. In the fifth edition of *Familiar Lectures*, this could mean deciding from as many as twenty genera and sixteen species, but most cases were far simpler—since many orders contained only a few genera with a few species regionally common, specific identification was often straightforward.

Even the most ardent advocates of the Linnaean system, however, admitted that it had major drawbacks. First, some plants were extremely difficult to classify because their floral parts were small, obscure, or not generally present. Second, the Linnaean system was rampant with strange bedfellows—including, for example, corn and oaks in the same class.[26] Both Lincoln Phelps and Eaton were familiar with other systems and freely admitted their scientific superiority, but they nevertheless held firm in their conviction that the Linnaean system was the best choice for teaching. As Eaton put it: "I would ask whether any botanist, who is in the habit of analyzing plants, however opposed to the Linnaean system, uses any other system for finding out the name of a plant, previously unknown to himself? Is it not a faithful guide on journeying to the Elysian fields of Vegetable Physiology? . . . The Natural Method is the grand *climacteric in Botanical science*, and the Artificial System is the brightest epoch of its youthful pupillage."[27] Through the many editions of their works the two authors continued to use the Linnaean system, adding short sections on natural systems of classification. Indeed, Eaton believed that the use of the Linnaean system in his manual "did more for the extension of the study than all the laborious research of every botanist in America."[28] Natural systems require a sophisticated understanding of morphology and physiology, and both Eaton and Lincoln Phelps felt strongly that this complexity precluded the use of natural systems by beginners.[29]

While the innovative teaching of Lincoln Phelps and Eaton at Rensselaer and Troy Female represented the best of botanical teaching in secondary schools during this first era, it was hardly the norm. Probably more common were schools that offered botany on demand or for an extra fee (usually about ten dollars) when a qualified instructor was available.[30] In those schools that regularly provided botanical instruction, courses ranged from a few lectures to actual work with plants. Further, botany was often offered informally, or hidden throughout the curriculum: common-school readers, for example, were rife with tales about natural theology that used plants as illustrations,[31] and teachers could and did explore the plant world with students of all levels without formal courses.[32] A few secondary schools and more colleges taught botany within courses on natural theology and evidences of Christianity. Medical schools, which required the study of materia medica, not only were a training ground for professional botanists (most notably Asa Gray), but also spawned many physician-botanizers.[33] Measuring the impact of such hidden botanical instruction is impossible, but it would seem that anyone who attended school for any length of time between 1820 and 1840 had some, though perhaps not much, exposure to the plant world.[34]

Gray and Wood

During the summer of 1835, yet another New Yorker set out to remedy what he saw as a deplorable lack of adequate texts on botany. This was Asa Gray, who had begun studying botany while a medical student and was soon in contact with John Torrey, America's leading botanist. Under Torrey's tutelage Gray had developed high scientific standards, and the recent texts of Amos Eaton and Almira Hart Lincoln Phelps did not impress him. Disappointed by Eaton's lack of the physiological and anatomical component necessary to enable students to use the field guide, he objected still more strenuously to the inaccuracy of Lincoln Phelps's texts. His criticism, however, went far deeper than accuracy or thoroughness: Gray split with Lincoln Phelps and Eaton over the philosophical issue of what system of classification to use. Gray, like his mentor John Torrey and unlike Lincoln Phelps and Eaton, was a botanist first and a purveyor of popular science second. To Gray and Torrey, who were far more familiar with the works of European botanists than were Eaton and Lincoln Phelps, natural classification was long overdue for adoption in the United States. While

Lincoln Phelps and Eaton addressed natural classification in appendices or separate sections, Gray was determined to incorporate it throughout his works.[35]

Underlying this philosophical split were two fundamental differences in the aims of Gray versus those of Lincoln Phelps and Eaton. First, Gray started from the assumption that the merit of botany as an academic subject lay not in such indirect benefits as mental training, but rather in the subject matter itself. The second difference, intimately related to the first, was that Gray wanted his writing to be of the same scientific quality as that of his European counterparts. While the Linnaean system had satisfied the pedagogical needs of Eaton and Lincoln Phelps, it fell far short of Gray's scientific standards.[36]

The publication of Asa Gray's intermediate-level *Elements of Botany* (1836) represented the dawn of a new era in American botanical education. It set new standards for botanical texts by including current physiological and morphological information, and by unequivocally endorsing natural classification. *Elements of Botany* was not nearly as successful in sales as Lincoln Phelps's or even Eaton's volumes, going out of print in less than a decade. Nonetheless, a diverse group of reviewers praised it warmly, and it converted some followers of the Linnaean system.[37]

In 1842 Gray, then at Harvard, improved on the out-of-print *Elements* with the first edition of the *Botanical Textbook*. In its early editions, the *Textbook* devoted 178 pages to "Structural and Physiological Botany" and only 113 to "Systematic Botany." This contrasted sharply with Eaton's total devotion to systematic botany, and with *Familiar Lectures*, which had 244 pages of systematic botany and a mere 76 pages of physiology and morphology.[38] Even more striking than Gray's increased emphasis on physiology and morphology was the content of his section on systematic botany: his treatment of the subject differed from Lincoln Phelps's and Eaton's in that it was not primarily a field guide, but rather a discussion of the principles of natural classification followed by a description of the orders, arranged naturally. Gray felt strongly that every student must come to understand individual plants as "organized and living beings" before undertaking the study of classification.[39] Students who wished to progress to fieldwork had to look elsewhere for a field guide.

The absence of a field guide in Gray's *Botanical Textbook* annoyed one educator so thoroughly that in 1845 he published a text of his own. Alphonso Wood, a Dartmouth-trained schoolmaster who readily admitted his amateur status in botany, was unsuccessful in his attempts to convince more-qualified botanists, including Gray, to write a book that combined up-to-date, basic botany with a natural flora.[40] In frustration, Wood himself wrote the *Class-Book of Botany*, combining a treatise on physiology and anatomy with a flora based on a natural system. An unsigned review in the *American Journal of Science*, a journal for which Gray regularly wrote, stressed the need for just such a book and concluded: "This work makes the study of plants interesting and fascinating, and must in our country supersede all the common works on the Linnaean methods. Teachers of academies, schools & c., will find it a noble work for their use in the study of plants."[41]

Wood had succeeded in producing a single-volume, reasonably priced, natural-system text aimed at the academy-level student. It was such a vast pedagogical improvement over existing options that it became widely used immediately. As table 7 documents, by 1850, 30 percent of New York academies and seminaries used the *Class-Book*, nearly driving Eaton out and cutting sharply into Lincoln Phelps's hold.

Despite the *Class-Book*'s pedagogical success, the scientific community viewed the work with mixed feelings. Gray and others readily admitted the *Class-Book*'s pedagogical superiority, but nonetheless vowed to mount "an uncompromising opposition" to its use.[42] Shortly after the first edition of the *Class-Book* appeared, Gray wrote to John Torrey: "Letters from [Edward] Hitchcock— and elsewhere—all point to the probability that they will have to use his [Wood's] book . . . and ask me to prevent it, by appending a brief description of New England or Northern plants to my 'Botanical Textbook.' "[43]

Gray and other scientifically oriented botanists held two reservations regarding Wood's book. First, Gray and Wood disagreed on the relative importance of physiology and taxonomy at the introductory level; the mere hundred pages that Wood devoted to physiology dismayed Gray. Even more offensive to Gray was Wood's flora. Wood had followed Torrey and Gray's *Flora of North America*—a path-breaking effort to provide a natural-system catalogue of North American plants—as far as it went, but it was not completely published and he had been forced to describe many families himself.[44] Gray feared that if the *Class-Book* remained the only single-volume, natural-system text, its taxonomic innovations would become entrenched. He asserted that his concern was not a loss of income from sales of his work, and he contemplated reducing the cost of the *Textbook* to make it cheaper than Wood's in the hope that, once challenged, Wood's errors would not become established.[45]

Gray decided eventually not to reduce the cost of the *Textbook*. Instead, still under the pressure of Wood's threatened capture of the textbook market, he strove to supply a series of alternative books that were both scientifically and pedagogically acceptable.[46] For several decades, both Wood and Gray continued to produce elementary, intermediate, and advanced texts, all of which went through multiple editions and printings.[47] Different locales did, however, develop preferences for one or the other, and over time Gray's works proved more enduring. In New York, Gray's texts did not surpass Wood's in use until 1870.[48] In the Midwest, schools used Gray's *How Plants Grow* (1858) "almost exclusively" in the years 1860–95.[49]

Wood and Gray shared an allegiance to the natural system that clearly distinguished them from Lincoln Phelps and Eaton. While the Linnaean system was no more than an index to assist in finding the name of a plant, the natural system of European botanists, derived from the work of Antoine Laurent de Jussieu and Augustin-Pyramus de Candolle, was an attempt to group similar plants together. The natural system and its underlying aims appealed to Wood and Gray for two reasons. First, unlike the artificial Linnaean system, it was not based on arbitrary characteristics, and hence they felt it was better science. Second, by revealing the similarities and differences of plants, it exhibited the relationships of plants in the plan of their Creator. While Gray realized that beginning students were primarily interested in identification, he argued that only by using the natural system would the student understand relationships between plants and thereby "gradually be led up to a higher point of view, from which he may take an intelligent survey of the whole general system of plants." This "whole general system of plants" and their relationships concerned Gray both scientifically and theologically. Wood and Gray both saw natural classification as God's plan or design in arranging plants and wanted their students to understand that plan.[50]

The popularity of Wood's and Gray's texts endured. As late as 1890, 94 percent of the reporting New York secondary schools used one or the other.[51] In 1895, over half the schools teaching botany in the Midwest still used Gray's *How Plants Grow*.[52] Despite the continued use of their texts, however, the era of Gray and Wood was drawing to a close. An obituary of Alphonso Wood in 1881 drew the distinction between the two clearly. The botanical community recalled Wood with friendly but qualified sentiments, praising his skill in writing clear and simple text but concluding: "As a scientific botanist his work can never rank very high, but as an educator his name will always be remembered."[53]

The division between pedagogical and scientific views of botany was, as we have seen, not a new debate, but in the closing decades of the century the two agendas diverged further as botanical education began moving in two separate directions: on the one hand, scientists pushed what they dubbed the "New Botany," which aimed to teach scientific thinking through laboratory instruction and rigor; on the other hand, educators promoting the "Nature Study" movement sought to teach nature appreciation through direct observation.[54] This split, which will be explored further in Chapters 9 and 10, was part of the growing distance between amateurs and professionals. Because of the paramount importance to amateurs of botany's place in education, however, it was a critical part; once the agendas of pedagogy and science went in separate directions, the strongest link between amateurs and professionals was lost.

5
Gender and Botany

There is no denying that botanizing held a special appeal for women and girls in the nineteenth century, not just in relation to other scientific hobbies but in absolute terms. From the beginning of its rise to popularity, women and girls were involved both in botanizing itself and in the movement to popularize the activity. Indeed, the appeal of botany to genteel women was so great that it came to be increasingly feminized. In 1822 Amos Eaton observed that the audiences for his public lectures throughout the Northeast were often heavily female, asserting that "more than half the botanists in New-England and New-York are ladies."[1] In Troy, where Eaton and Almira Hart Lincoln Phelps collaborated to introduce botany into the secondary-school curriculum, there was sufficient interest among the public by the late thirties to warrant an all-female class taught by a woman three days a week.[2]

By mid-century, botanizing was so popular among women and

girls that some critics found it necessary to refute charges that it was too effeminate an activity for boys and men. In 1837 one author wrote that "the cultivation and study of flowers appears more suited to females than to man."[3] By the 1850s botany had become so clearly identified as a ladylike pastime that one editorial rued the fact that "the boy who, having an eye to see and a heart to feel the beautiful in nature, undertakes to master the charming science is taunted as a 'girl-boy' and as unmanly," and Emily Dickinson, in a poem, addressed a botanist as "she."[4] As late as 1887 botany was still so closely associated with women and girls that *Science* ran an article entitled "Is Botany a Suitable Study for Young Men?" which sought to assure boys that botany was not "merely one of the ornamental branches suitable enough for young ladies and effeminate youths," but could be respectably undertaken by "able-bodied and vigorous brained young men."[5]

To say that botany was immensely popular among nineteenth-century American women and girls is not, however, to say that women's participation was the same as men's, or that girls botanized for the same reasons and in the same ways that boys did. Indeed, there were major differences that offer much insight into gender roles in middle-class culture of the nineteenth century, as well as about what it meant to botanize.

Early Attitudes

The reputation of botany as a science more conducive than most to female participation predates the American experience. In Britain, female participation in botany was legion in the late eighteenth century, and special texts, which desexualized the Linnaean system for female audiences, became the order of the day. In France, Jean-Jacques Rousseau's 1771 *Essais élémentaires sur la botanique* (published later in English as *Elements of Botany Addressed to a Lady*) was written as an aid for a woman who desired to guide her daughter in the study of botany. These volumes, and the tradition accompanying them, made their way to America and were transformed by the new environment.[6]

Historians of science in America have taken much interest, for example, in Jane Colden, whose scientist-politician father taught her the Linnaean system and introduced her to the world of scientific correspondence as a means of alleviating her boredom with rural living. Cadwallader Colden introduced Jane to botany because "I thought that botany is an amusement which may be made agreeable to the ladies, who are often at a loss to fill up their time. Their natural curiosity, and the pleasure they take in the beauty and variety of dress, seems to fit them for it."[7] Jane Colden became a novelty both for her contemporaries and for historians who have looked in desperation for evidence of women pursuing intellectual activities. Focusing on Jane Colden, historian Joan Hoff Wilson borrowed Samuel Johnson's remark to James Boswell—"a woman's preaching is like a dog's walking on his hinder legs. It is not done well; but you are surprised to find it done at all"—to describe the early fascination with female scientific interest, especially in botany, during the Colonial era.[8]

While we do not actually know much about how Colden went about botanizing, the indications are that she did so in a way that became the pattern for later American women—a pattern of activities very different from that of men. Accounts of Colden's botanizing invariably mention the gardens at Coldengham, the family estate, and also the specimens sent to her by correspondents. Missing, however, is any discussion of field trips and exploration.[9] That should not surprise us; after all, she was female. It does suggest, however, that her botanizing differed from that of, say, William Bartram, her contemporary and correspondent, who spent extended periods in the wild collecting specimens. And it ought to cause us to ask how gender affects the practice of science, even at "lower" levels.

Indications that women would prove freer to botanize than to do many other things also began early. As we have seen, public lectures on botany in various parts of the Northeast offered by Amos Eaton and others were attended by large numbers of women, and (due in no small measure to Eaton's influence) botany also became a common part of the curriculum as female seminaries spread during the 1820s and 1830s.[10] The reasons for the popularity of botanizing were many. Most notably, botany was considered the most genteel and delicate of the sciences.

Nineteenth-century observers commented especially enthusiastically on the special attractions that gentility had for female botanizers. Women, especially those of the middle class, strove at least in their rhetoric to achieve an idealistic virtuous state, dubbed "true womanhood," that embodied piety, purity, submissiveness, and domesticity.[11] Botanizing easily fit the mold of acceptable activities for the true woman. First, collecting and studying plants were acceptably genteel, as popular essayist Wilson Flagg perceived: "Women cannot conveniently become hunters or anglers, nor can they without some eccentricity of conduct follow birds and quadrupeds to the woods . . . the only part of natural history which they can pursue out of doors is the study of plants."[12] In addition, the subject matter itself, the plant world, was sufficiently refined for female attention. Flowers, with their "fragility, beauty, and perishable nature," were likened to "a pure-minded and delicate woman, who shrinks even from the breath of contamination."[13] Surely plants could in no way be considered crude or unrespectable. One author expressed the hope that botanizing might provide ladies, especially those of the South, with an escape from such trifling and pernicious habits as "lolling on the sofa and reading a trashy novel."[14] These appeals made botany an ideal addition to female activities and curricula at a time when middle-class moralists were striving to prescribe a well-defined, narrow sphere in which women might properly function.[15]

At the same time, botany appealed to those who were challenging the notion that women belonged in a separate sphere from men—or, less radically, were striving to expand women's sphere. Botanizing may have fit the prescriptions for true womanhood but it was not restricted solely to females, and some of its activities went well beyond the conventional bounds of female behavior. Those, like Almira Lincoln Phelps, who sought to challenge subtly the limits on female behavior prescribed by true womanhood, found botany a useful tool because of this range of possibilities. Once the more innocent aspects of botanizing became acceptable female behavior, the more controversial activities could be slowly added without attracting undue notice.[16]

While gentility made botany "a science particularly adapted for ladies," many males—contrary to the critics' deepest fears—also found it an attractive hobby. In some cases, boys and men even admitted being attracted, not repelled, by the genteel nature of botanizing.[17] Regardless of gender, those who possessed or coveted a "tender and delicate mind" found that "in botany all is elegance and delight."[18] Indeed, one author argued that "all persons of education, of both sexes, either have obtained, or wish to acquire,

The PASSION FLOWER *Belongs to the class*
MONADELPHIA *and order* PENTANDRIA.

See Page 165.

"The Passion Flower" (Frontispiece from J. L. Comstock, *The Young Botanist*, 1835)

some knowledge of this fascinating science."[19] Just as it was an acceptable alternative to the trifling hobbies of young ladies, so it was a method of diverting young men from spending their time at deleterious pastimes such as trotting matches, "where the seeds of vice are first sown, and habits of dissipation formed."[20] Botany could clearly complement or enhance a genteel lifestyle regardless of gender. Male or female, nineteenth-century Americans found botany a suitably genteel study.[21]

The gentle exercise of collecting appealed greatly to a society concerned about female health, especially that of schoolgirls. To antebellum eyes, the combination of the sedentary life and the fragility traditionally attributed to females put schoolgirls at special risk for a host of ailments, including "headaches, languors, sleeplessness, indigestion, and a thousand other ills," which exercise would prevent.[22] Educators and doctors alike recommended daily walking, as much as two to three hours' worth per day, suggesting that an outdoor hobby, such as botany, would lend interest and encouragement to an otherwise dull activity.[23] Almira Lincoln Phelps saw this as of special importance to women: "The study of Botany seems particularly adapted to females, . . . its pursuits leading to exercise in the open air are conducive to health and cheerfulness. It is not a sedentary study which can be acquired in the library, but the objects of the science are scattered over the surface of the earth, along the banks of the winding brooks, on the borders of precipices, the sides of mountains, and the depths of the forest."[24]

While stressing women's alleged frailty and delicacy, and reflecting the growing nineteenth-century concern over female invalidism, reformers argued for an extension of women's activities in the name of fitness.[25] Hence, searching for botanical specimens became a supplement to other, more formal exercise (especially calisthenics) for students at Troy, Patapsco, Mount Holyoke, and other female schools.[26] As Phelps's description suggests, however, the exercise of botany sometimes exceeded the gentle strolls that were ideal. Indeed, the vigorous side of botany caused a fair amount of controversy: just how much exercise is genteel, especially for women? Even those like Wilson Flagg who championed female involvement in botany had reservations: "Even in this field [botany] they meet with obstacles not encountered by the other sex. A Young Lady cannot safely roam at will in any place at any distance. . . . [W]hile a young man may traverse the whole country in his researches, his sister must confine her walks to the vicinity of her own home and to the open fields and waysides, and in those limited excursions she sometimes needs protection."[27]

The parameters that Flagg prescribed reflected his views of what genteel women and girls ought and ought not to do, and were based on his recollections of botanizing with his sisters during his antebellum childhood. On the one hand, botany was the ideal area of natural history for a genteel woman to pursue. On the other hand, a genteel woman should not remove herself from the safety of women's traditional sphere. Hence, even mildly rigorous collecting required male protection in order to preserve decorum. Flagg's position was by no means unusual. Fanny Alexander, who as an adult wrote extensively for other botanizers, recalled being punished frequently as a youngster for collecting without proper accompaniment.[28] While a botanizer who was concerned with advancing science would have found Flagg's prescription restrictive, most amateurs seem to have been less concerned with the science itself than with their own self-improvement and recreation. Flagg's restrictions on the freedom of women to collect need not necessarily have interfered with either. Nevertheless, Flagg made it quite clear that in his family they had: "My own interest in botany was first awakened by collecting flowers for my sisters which they afterwards analyzed and named."[29]

Men and boys frequently collected for women and girls. Even in fictional accounts, "the boys chose the wet and rugged places, leaving to the little girls, the smooth and dry."[30] One fictional girl rued the fact that in order to obtain a choice specimen she had gotten her feet wet, a transgression severely warned against as being one of the few health hazards of botanizing.[31] Lincoln Phelps herself mentions the pattern of boys collecting for girls, in describing activity at Troy: "The region around Troy is rich in its flora, and scarcely a dell, ravine, or island of the Hudson in its vicinity, was not explored in these expeditions. The young gentlemen students of the Rensselaer School were chivalric and indefatigable in their efforts to produce specimens for the ladies' herbaria and so Botany became the fashion of the day."[32] Lincoln Phelps went on to describe a classroom scene of "young girls, radiant with bright eyes and glowing cheeks," each keying out her share of the specimens the young men had collected, implying that the female students did most of their botany indoors. At Lincoln Phelps's Patapsco Female Institute, which she ran from 1841 through 1856, students seem to have operated under similar constraints; given that they could not leave the campus even to attend church services, we must question the degree to which "botanizing in the woods" can really have been a regular student pastime, as the 1843 catalogue reported.[33]

Lincoln Phelps's schools were by no means anomalous. Her botanical mentor Amos Eaton described guidelines for hiring a man to collect specimens for a class, but cautioned that "all gentlemen students should collect their own plants in the field."[34] Accounts of classes often included the acknowledgment that those specimens that the women or girls collected themselves were supplemented by flowers gathered by male students or teachers from "wet and inaccessible places where they could not go."[35] Or listen to the advice of Mrs. S. E. Roessler, who studied botany under Lydia Shattuck at Mount Holyoke in 1851–52:

> I suppose many of the young readers of the *Observer* are pursuing the study of botany. I wish to whisper to the girls in the botanical circles to keep on the right side of the big boys. Now do not say they are "too fresh" or even suggest that they are between "hay and grass." These same boys will scale rocks and jump from bog to bog of the treacherous swamp to secure rare plants for you. They will bring you whole bunches of whip-poor-will shoes and side saddle flowers also with pitcher-like leafage that rank among curios of vegetable growths. Now do not doubt my word, for I have been there. But this is *sub rosa*.[36]

Mount Holyoke students did far more of their own collecting than those at Troy. At the time Roessler studied there, students had to identify between three and four hundred species during the two-year sequence of botany (a total of about twenty weeks). This requirement led the college president, herself a botanizer, to ban picking flowers for ornament or taking multiple specimens for pressing within three miles of the campus, because overpicking was threatening several species. One suspects, however, that male students from nearby Amherst may have been as obliging as those of Rensselaer.[37]

Behind this pattern of males collecting for females lies another probable cause of the popularity of botanizing among the young: its social aspect. As an eight-year-old Southern girl observed in a letter to her parents in 1833, reporting that she had become sufficiently proficient at botany that the family tutor had decided to instruct her with her older brothers, "We like it better than being in a class by ourselves."[38] The possibilities for socializing while collecting in groups or in collecting for a friend or beau were great and certainly added to the appeal. At Antioch College in the 1860s, for example, certain parts of the surrounding countryside were restricted to men on odd days and women on even days—but anyone could botanize on any day.[39] This rule may well have increased the popularity of botanizing dramatically!

Underlying the conflict between gentility and exercise was a tension over shifting roles for women. On the one hand, enthusiasts throughout the century hailed botany as genteel. On the other hand, botanizing could exceed the traditional limits on genteel female behavior, if practiced fully. This is seen clearly in a short story in the *Water Cure Journal* in 1855, in which a young woman discusses the response of a new male acquaintance to accompanying her botanizing, an activity he thinks best left to men because it "rob[s] a woman of her femininity." They start at odds. She finds his use of tobacco offensive, and he reacts unfavorably to her Bloomer outfit. "Now how could I, with women's ordinary dress, ever scale that five-rail fence, cross that ravine, ford that stream, climb that hill, walk yonder prairie, or ramble through those old woods. . . . But I can do it easily dressed as I am now." During the course of the afternoon's collecting, the power of nature exerts itself: "We met as strangers, we part as friends; you have promised to give up tobacco and advocate freedom for women, even in dress, and thus shall man be free."[40]

Changing Mores

As the century progressed, the parameters of acceptable female behavior changed. Middle-class women, for example, became increasingly involved in reform movements, and many entered the teaching profession. These moves away from the home slowly, yet dramatically, expanded women's sphere—a change also reflected in attitudes toward collecting. The result was confusion over just how much collecting was proper. Because of this tension, the degree to which girls and women relied upon boys and men varied. Some, like Wilson Flagg's sisters, appear to have been largely dependent, because they clung especially closely to the code of gentility. Others, like Lincoln Phelps's Troy students, balancing the code of gentility with the desire for exercise, often enlisted the aid of men and boys to do the difficult or messy work. As the antebellum period drew to a close, other advantages of botanizing, especially physical culture, eclipsed gentility in attracting women to the activity. In the post–Civil War era the diversity of experience became greater, and many women felt far freer to collect.

One of the places where this change and its effect on botanizing are revealed most clearly is in fiction, especially in fiction aimed at women and children. Early in the century, Americans reprinted British didactic volumes, like Priscilla Wakefield's *An Introduction to Botany*, in which a young woman writes about her botany lessons in a series of letters to her sister, who has gone to visit an aunt: "My kind Mother, ever, attentive to my happiness, concurs with my governess in checking this depression of spirits, and insists upon my having recourse to some interesting employment, that shall amuse me, and pass away the time, while you are absent. My fondness for flowers had induced my mother to propose Botany, as she thinks it will be beneficial to my health, as well as agreeable, by exciting me to use more air and exercise than I should do without such a motive."[41] And so the volume continues, with the stay-at-home sister writing often to share what she has learned in her botany lessons, both of botany and of life. This format of lessons disguised as fiction was a common one, designed to make learning more appealing.

Antebellum American attempts at the genre reflect a world in which boys did the messy work, and girls admired and analyzed the flowers while botanizing just enough to get some exercise (if they botanized at all). In the 1870s and 1880s, however, a new flurry of fictional botanizing emerged, with a wider range of behaviors portrayed. Botanizing was more consistently shown as a social activity, often combined with other activities—for example, photography, or making an autograph album as a souvenir of a vacation—or was suggested as a topic shy children could successfully employ in conversation with adults.[42] More interesting, perhaps, is the explicit portrayal of appropriate gender roles. In this later fiction girls are much freer to collect for themselves. Louisa May Alcott's Rose, for example, spends her vacation with her cousin Mac in the mountains. Together they roam at large, he geologizing, she botanizing, and teach each other their respective specialties.[43] In one of the last didactic volumes in the old mold, three children—Edie, Clara, and Malcolm—study botany together under the instruction of their governess, Miss Harson. Malcolm at one point is appointed to be the leader on a collecting trip taken without Miss Harson, but he gets them all lost and thus loses what little authority his gender has afforded him. He is also no better nor worse a student than his sisters. In short, the three children are equals in their botanizing.[44]

Further reflecting these changes, one fictional account from 1885 told of a coeducational group of students who took a week-long collecting trip with their instructor and his wife. While the boys gathered geological and zoological specimens, the girls collected plants. When it came to the difficult specimens—for example, water lilies in the middle of a lake—the boys lent a hand, but clearly the story reflects a much freer environment, a bigger sphere, than Lincoln Phelps's students experienced in the Troy classrooms of the twenties or the Patapsco classrooms of the fifties.[45] This freer atmosphere was not merely fictional or prescriptive. While botanizing was still touted as good preparation for other genteel accomplishments such as creating wax flowers and painting, writing travel memoirs, and making polite conversation, many women were less concerned than their antebellum predecessors about traditional standards of gentility and collected their own material while maintaining their respectability intact to varying degrees.[46]

In Syracuse, Philadelphia, Wilmington, and other towns, especially those where women were unwelcome in men's clubs, women organized botanical clubs to do fieldwork and indoor work together for camaraderie and to share what they learned. The Syracuse Botanical Club, founded in 1878 for mutual instruction and "to induce women, particularly, to occupy themselves in a way both improving and pleasant," organized weekly excursions during the collecting season, maintained a library and herbarium, and carried on—collectively and individually—correspondence with a variety of botanists elsewhere, including the Harvard group. They successfully established a number of new stations of known species and catalogued the local flora in the process. Club activities were supported by fund-raisers, including card parties, and fierce rivalries over who had found what first erupted frequently.[47]

Some women became botanical helpmates for husbands, fathers, and brothers, going on botanizing honeymoons and on collecting trips as assistants. J. G. Lemmon and his new wife, who had met through their botanical interests, honeymooned by taking a collecting trip. The *Botanical Gazette* commented on one of their journeys as follows: "Mr. J. G. Lemmon and wife are off again for Arizona. One would have thought that the experience of their last trip would have sufficed for a lifetime; but as long as a plant remains to be discovered, these intrepid explorers will try to find it."[48] J. Reverchon described the decision to end a two-and-a-half-month collecting trip with his wife, and a hand, through the wilds of Texas in 1885: "Thus far we had had a tolerably pleasant time, in spite of gnats, mosquitoes and other insects, but the dry weather had now set in, the heat was increasing alarmingly, the water was sinking very fast into the sandy beds of the rivers, and what was more important to me, the vegetation was beginning to shrivel up and disappear. Our team was jaded, our provisions consumed, our clothes in tatters, our finances exhausted. We had either to refit our expedition or retreat, hence after consultation, the march on Mexico was postponed and a retreat ordered."[49]

At seminaries and colleges and in private classes, female students and their teachers gathered at least some of their own specimens, some of the time. Annie Oakes Huntington, who instructed her fellow Boston society ladies, wrote often about her exploits to her friend Annie F. Rogers: "I have become a Bedouin since my last letter, and I'm never going to live within walls again."[50] And increasingly, women were able to obtain teaching and administrative positions in botany.[51]

Mary Katherine Brandegee, for example, took up the study of botany as a young doctor in the late seventies when her struggling practice failed to keep her sufficiently occupied. Within a very few years she gave up her practice altogether and became the curator of botany at the California Academy of Sciences. Botanizing extensively—first on her own, and later with her second husband, Townshend Stith Brandegee—she added dramatically to the nascent Academy's collections. Through her influence the *Bulletins of the California Academy of Sciences*, *Zoe*, and the Botanical Club (a

California institution) were all founded.[52] And through her aid many young botanists (many of them female) got their start—most notably, Alice Eastwood.

Alice Eastwood began her first serious collection of botanical specimens the summer after her graduation from high school in 1879, collecting at first to amuse a group of small children in her charge, and then for herself. Self-taught, she worked as a school-teacher, scrimping through the year to support botanizing trips during the summer to Cambridge, where she met Asa Gray, and to California, where she met the Brandegees. In 1892, at thirty-three, she moved to San Francisco at the urging of Katherine Brandegee, who in effect gave up her job in order to offer it to Eastwood so that she would be able to come. She remained at the California Academy of Sciences as its curator of botany and as editor of its botanical journal, *Zoe*, for fifty-seven years, markedly adding to the Academy's collections with her own specimens gathered on innumerable expeditions, and heroically saving a large portion of the collection from fire following the 1906 earth-quake.[53]

While Alice Eastwood's bold behavior certainly shocked some, she was by no means the extreme. Other women rejected propriety and pursued botany freely at the expense of gentility. One of the most striking examples was Kate Furbish of Brunswick, Maine, who botanized and painted Maine wildflowers from her childhood in the 1840s until her death in 1931. Furbish's letters to botanists elsewhere provide an astounding account of her collecting. In one she apologized for the condition of some specimens, explaining that she had been unable to carry a vasculum up the mountain on which they were collected, as the climb was difficult, with no trail in many places.[54] Reflecting on her exploits, she remarked: "I have wandered alone for the most part, on the highways and in the hedges, on foot, in Hayracks, on country mail stages (often in Aroostook County with a revolver on the seat), on improvised rafts, . . . in row-boats, on logs, crawling on hands and knees on the surface of bogs, and backing out, when I dared not walk, in order to procure a coveted treasure. Called 'crazy,' a 'fool'—and this is the way that my work has been done."[55]

Furbish cared little about what others thought of her lack of decorum, worrying rather about what was thought of the quality of her specimens. This boldness sometimes provoked concern and censure. She, herself, reported that she often had to apologize for anxiety she caused in others who worried when she stayed out too long collecting—for example, when she was delayed because she had seen "some water plants which I needed and it took me three hours to build a raft."[56] Certainly this would have been unthinkable behavior for a woman a half-century earlier.

As women's sphere expanded in other areas, so it expanded in botany. Women were progressively able to do more fieldwork. Gentility attracted many early- and mid-nineteenth-century women and girls to botany. For some of them, botanizing became an opportunity to extend women's sphere subtly without directly challenging dominant notions of gentility. Yet even so subtle and modest a threat to tradition as collecting botanical specimens aroused resistance. Ironically, the same code of genteelness that drew women to botany also limited their ability to pursue it fully. Like education, abolition, and the host of other arenas that women entered during mid-century, botanizing opened up to women because it was consistent with traditionally accepted female behavior. Moreover, as women pursued botany, they expanded the range of acceptable female behavior. By century's end, gentility was less of a hindrance to female botanizing, but it was also less of a drawing card to the science "particularly adapted for ladies."

6

Botanizing and the Invention of Leisure

During the nineteenth century, American attitudes about work and leisure changed, and botanizing evolved from a work surrogate into a recreation. Antebellum Americans, especially "middling, largely Protestant, property owning" Northerners, worried a great deal about filling nonwork hours productively.[1] Reflecting this concern, promoters portrayed botanizing as worklike, stressing the purposeful physical and mental exercise that it involved, as well as the disciplined, regular habits required to do it well. These worklike attributes made it possible for botanizing to serve as a proper substitute for work, thereby countering the sloth and idleness that moralists so loathed.

By mid-century, the nature of work was rapidly changing, however, first in the Northeast and then elsewhere, as the new nation became industrialized. When work moved from the farm to the factory, ideas about the value of both work and leisure changed.

Americans learned how to play, becoming more interested in relaxation and outdoor exercise.[2] Worklike though it was, botanizing was also pleasant enough to attract those looking for a recreation to fill their leisure time. Botanizing remained widely popular throughout the nineteenth century by satisfying both the traditional attraction of work and the new interest in play. Botanizers balanced their hobby between work and play, enjoying the best of both ideals.

The Work Ethic

The importance that early Americans assigned to work ran throughout their culture. Middle-class Northerners held such a firm belief in the moral primacy of work that they sometimes resorted to extreme activities in their search for work surrogates, such as setting children to dull toil for the purpose of teaching them self-discipline and good work habits. One especially zealous promoter of the moral value of work was so concerned about avoiding idleness that he regularly shoveled sand from one corner of his basement to another.[3] Not all botanizers were Northerners or members of the middle class, but because many promoters of botanizing were, these Northern, middle-class ideas about work pervaded the botanical literature.

Nineteenth-century Americans had no difficulty construing botany as worklike: like work, and unlike play, it involved the disciplined expenditure of energy toward a productive end. Listen, for example, to the words of Alphonso Wood, an academy teacher charged with keeping youths from idleness and teaching them the value of work: "Let it not be said that botany attracts such willing votaries because it requires no labor, no persevering effort. No science is more intricate or profound. It cannot be understood except by vigorous and persevering effort."[4] Botanizers saw the collection, identification, and preservation of botanical specimens as labor with a purpose, labor that required regular expenditure of time and energy if it were to be done well. Botanizing converted idle roaming out-of-doors into self-improvement, and the admiration of flowers into science. Enthusiasts proudly detailed the work involved in each step of the process, boasting often that theirs was no easy task.

Those who wrote for botanizers urged their audience to devote as much energy as possible to the science in order to derive the greatest benefits. As one critic remarked, after extolling the virtues of botanizing: "It is true that the superficial study of this, as of any other science, will be of little use."[5] Botanizers were to work not simply during those seasons when one could collect, but in winter as well, studying summer specimens in preparation for the upcoming season, or helping a less experienced botanizer to prepare for fieldwork. They were encouraged to collect their own specimens, rather than purchasing sets, and to collect vigorously enough to have extras to exchange with other enthusiasts. Writers urged botanizers to spend rainy days and evenings year-round at indoor work so that fine days and daylight were free for fieldwork. Botanizers were, in the words of one late-century advocate, to "ride it early and hard."[6] The message that only work could make botany worthwhile pervaded the botanical literature, forming the basis of many a textbook introduction and essay.

Botanizers who took their hobby seriously had, of course, real reason to consider their activities as work. Collecting journeys could, and often did, involve a good deal of exertion. Properly preserving large numbers of specimens could take hours. Describing specimens, a necessary step for the serious botanizer, required patient and careful mental labor. Identification involved not only book work, but also extensive correspondence, as did the exchange of specimens with other enthusiasts.

At the same time, botanizing was also fun and amusing, characteristic of play. Walking out-of-doors and looking at flowers were pleasant. Until late in the century, however, work prevailed as the moral ideal to which botanizers compared botanizing. Nowhere was this more so than in the literature advocating botanizing as an appropriate pastime for children. Those who wrote for and about children saw botanizing as a means of instilling good work habits early on. Some of the clearest examples of how botanizing encouraged diligence, and other traits of good workers, came in fiction written for children, most notably Jacob Abbott's *Rollo's Museum*.

Jacob Abbott, a Congregational minister who had been a tutor of mathematics and natural philosophy, wrote more than a hundred books for children between 1829 and his death in 1873. Many of his works were exceptionally popular, but the most widely read were the two dozen didactic tales about Rollo, a New England farm boy. In *Rollo's Museum*, written in 1839 and revised and reprinted through the seventies, Abbott made it quite clear that in order to be worthwhile, natural history, specifically botany, had to be hard work as well as fun. The industrious and curious Rollo started his museum to fill a period during which his physician had ordered him to do no reading. Rollo and his friends pursued many areas of natural history, but it was botany to which they brought the most intensity, organization, and interest. At first the children randomly picked and identified plants, with no regularity or diligence. Soon, however, they began carefully to collect, preserve, arrange, and identify plants. Later they bound their specimens into books, which not only served as references for new finds, but also pro-vided a source of entertainment year-round.

Abbott's characters found botany more interesting than other areas of natural history, and it was botany at which they worked hard enough to be rewarded by learning and a pride in achieve-ment. Rollo's museum became more than a jumble of objects only when Rollo and his friends began to work routinely and pur-posefully at botany, rather than simply collecting interesting ob-jects haphazardly. Implicit in Abbott's story was the message that natural history, including botany, became a worthwhile or even pleasant pastime only when it was pursued as one would pursue work, with diligence, regularity, and purpose. One could then take pleasure not simply in the flowers or curiosities at hand, but also, and more importantly, in the satisfaction that inevitably followed a job well done.[7]

The theme of children working at botany abounds in antebellum fiction. Fictional children often found botanizing treks more fun than regular walks. They begged for excursions, and they volun-tarily put aside play when an adult in their midst was willing to talk about plants. In a typical story from the forties, a female botanizer on a walk encountered a group of children gathering flowers. Two boys "with earnest gaze, and brows on which thought was deeply impressed," looked beyond gaudy flowers at mosses, much to the amusement of their frivolous companions, who were gathering bouquets of pretty wildflowers. The woman, who conveniently was carrying a dissecting kit, hand lens, and field guide, delighted the young moss collectors by making prompt identifications. She then walked with them to their home in order to ask their parents whether the boys could come to see her collections. The woman

"WHAT YE GOT NOW?"

"What Ye Got Now?" (From Amanda B. Harris, *Wild Flowers and Where They Grow*, 1882, 71)

soothed the mother's complaints about having to clean up after the specimens were studied by explaining that botanizing was teaching the boys how to work. Both parents agreed and acknowledged that natural history was innocent and potentially useful, one of the few such activities the family could afford to pursue.[8]

The young naturalists of this story found botanizing more appealing than idle play, while the adults applauded their diligence and purposefulness, adding that the activity was also innocent, useful, and inexpensive. Thus while the promoters of botanizing often contrasted it with play, some more perceptive enthusiasts realized that the pleasant aspects of the activity were a drawing card, especially for children. This is not, however, a tale of children choosing botanizing over a conventional form of children's work such as school tasks, farm chores, or sewing—it is a story of children choosing botanizing over play. These boys and their counterparts enjoyed the purposeful work of botanizing more than the play of flower gathering, a trait their adult creators applauded.

Not all fictional children, however, appreciated or enjoyed the hard work involved in botanizing. In an 1833 fictional encounter between Maria, a city girl, and her country cousin, Maria expressed her delight at the prospect of promised botany lessons. Her cousin, who shirked all work, remarked that she had abandoned her study of botany because "there are so many hard names and it takes forever to learn how to analyze a flower." Maria quickly saw through her cousin's complaint to the underlying laziness. Maria's conclusion was that botanizing required work but was not too hard for her.[9] Implied in the story was a defense of botanizing from charges that it was too much work and too difficult for children. Such defenses were a nearly universal component of botanizing texts through the fifties and persisted, though less frequently, for decades.[10]

Educator Almira Lincoln Phelps firmly believed that botany involved work, but she did more than any other individual to dispel the notion that it was too hard for children. Her popular text *Botany for Beginners* began, "You are now about to commence a study which was formerly thought too difficult for children, but which is, in reality, much easier than many which they usually attend." Botany was not, Lincoln Phelps insisted, "dry and difficult" unless improperly taught. Throughout her writings she remained firm: Botany is work, and hence a worthwhile activity, but not excessively difficult work.[11]

In defending botany from the reader's hypothetical objection that it was unsuitably dull or difficult for children, an anonymous enthusiast wrote in 1831: "Of the fallacy of these statements I am fully convinced, . . . having seldom known children unwilling to relinquish less intellectual sports for a walk to collect specimens of this nature, with a view to the pleasure of investigating them."[12]

Also writing in 1831, Almira Lincoln Phelps echoed these senti-ments, arguing that "even children may become botanists, and lay aside their toys to divert themselves by distinguishing the organs of plants and tracing out their classification."[13]

It should be noted that sentiments like these, while reflecting the pleasure children took in botany, probably reveal more about the adult observers than about children. Lincoln Phelps and the author of *The Pastime of Learning* shared a concern with most other antebellum, middle-class adults about filling children's time usefully and innocently. The arbiters of values rejected the con-cept that fun and pleasure should be criteria for selecting chil-dren's activities. Phelps and others did, however, realize that useful activities that were also fun and pleasant would appeal to children, an appeal that could be used to great advantage. Reflect-ing this practicality and perhaps also a more perceptive view of child nature, Samuel Goodrich, author of countless pious and pedagogical works, suggested that children keep a "folium" or scrapbook of pressed plants. This would, he suggested, be a useful and healthful pastime for children. Some children would love it "almost as well as they do their play."[14]

The willingness of children to botanize and the evident pleasure they took in it attracted the attention of those who wrote texts. Botanist and textbook author Asa Gray, for example, argued in an 1842 lecture that children learn more on their own than they do at school. In light of this, Gray stressed the importance of providing children with books on natural history and cabinets for systematic study. Children would then teach themselves natural history with great benefit to their intellectual development, self-discipline, and character. Like Rollo, Gray's children were to select botany to fill their free time.[15]

The hopes of Lincoln Phelps, Goodrich, and Gray were realized at least in part. Real children of the mid-nineteenth century expressed enthusiasm for botanizing and often chose it over other activities. Autobiographical and biographical accounts suggest that botanizing was an extremely popular childhood pastime.[16] Moreover, children frequently wrote letters about botanizing to the periodicals that catered to them. Typical was a letter in 1847 to the *Youth's Cabinet* from female students of the Newpark School in Lewisberry, Pennsylvania. After describing the town they lived in, they turned their attention to the magazine. "We like it very well," they began, "but we think we would like it still better if you would give us more of Natural History and Botany in its pages."[17] In their letters, children demanded botany as well as or instead of fiction, puzzles, and other diversions. Their periodicals responded by giving them essays on plants, puzzles with botanical themes, hints on botanizing, and especially fiction with botanical themes.

The Pleasure of Work

Asa Gray and Samuel Goodrich saw the pleasant and enjoyable nature of botanizing as a means of attracting children to the science, but not as a reason in and of itself for botanizing. Americans did not initially accept pleasure alone as a legitimate motivation for botanizing: certainly botanizers shaped their activities to make them as pleasant and interesting as possible, but pleasure was secondary, an added bonus. By mid-century, however, there were hints that the recreational aspects of botanizing were becoming more and more important. Americans adjusted to industrialization by learning to play. While the work pace of farms and homes fluctuated between busy periods and slow spells, the pace of industrial work was constantly fast.[18] The promoters of botany devoted new attention to the pleasure of botanizing, stressing that botanizing could be *both* worklike and fun.

One example of the increased emphasis on balancing pleasure and work was the new ardor for collecting one's own specimens. Early advocates, including Amos Eaton and Almira Hart Lincoln Phelps, had advanced the notion that learning from specimens was more interesting than studying books. They and others soon argued that personally collected specimens were far superior to materials provided for the botanizer by someone else. One enthusiast of collecting stated that botanizers would find working on their own specimens "infinitely more amusing, interesting and satisfactory than looking at specimens purchased ready mounted and labelled."[19]

The comments of Wilson Flagg, a popular lecturer and essayist on natural history, illustrate that far more was involved in the rise of collecting than just exercise and amusement. As a boy, Flagg often collected for his sisters, whose botanizing was confined to the comfort and safety of their home. He had felt even then that he "enjoyed the principal pleasure of the pursuit, while they performed all the drudgery." Flagg went on to extol the virtues of the risk and adventure of collecting, and to assert that the specimens he valued most were those he had worked hardest to obtain. His sisters were able to share some of the adventures by his recounting them, and he recalled that "any time I had got a ducking, or had come home covered with mud, or with bruised limbs, or a scratched face, in my scrambling after a rare plant, my mishaps gave it [the specimen] in their eyes an additional value." Flagg's sisters valued his efforts and appreciated the rarity of some of the specimens he brought them. The amount of work involved served to measure the value of his collecting. "There is a philosophy in these matters, which has never yet received the attention it deserves, and is still very imperfectly understood, especially by those who would make the path through every field of learning so smooth and easy as to excite nothing of the spirit of adventure."[20]

There is no doubt that botanizing was a pleasure to young Wilson Flagg. It is equally clear that his pleasure was directly related to how hard he worked at it. The work of botany was what made it a pleasure for Flagg and for others, including Alphonso Wood. Wood, academy teacher and textbook author, devoted an 1860 essay to the necessity of working hard at botany in order to derive any benefit or understanding from it. He prefaced the essay by describing the satisfaction felt after a vigorous collecting trip and a long evening's work with specimens, "toiling for hours as no schoolmaster could have compelled us to do, being attracted to the task by the very love of it alone." Wood went on to describe, in terms quite similar to Wilson Flagg's, the direct correlation of work and pleasure in botanizing.[21]

Wood and Flagg took pleasure not simply from exercising outdoors or from learning directly from nature, but also from the toil involved. The work of botanizing was its own reward. Carrying this sentiment to its logical extreme, many enthusiasts rejected the use of domesticated plants, not because they were so modified as to be freaks, but because they made botanizing too easy. Flagg put it succinctly: "Botanizing in a garden is like gunning in a poultry-yard."[22]

Taking the work out of botanizing by using domesticated specimens rendered it dull. Removing the work of analysis and identification of specimens left collecting purposeless, and thus approached idleness. Gathering flowers for pleasure, for example, was idleness. One supporter of botanizing attacked the gathering of flowers for May baskets as having "no valuable qualities whatever" and as having originated among "the lower classes, and the more ignorant and down-trodden portions of the people of Europe," a group the author found no reason to emulate.[23] The purposeful work of botanizing rescued it from the class of idle amusement, transforming mere play into an acceptable activity. In order to make botanizing worthwhile, one had, according to a host of enthusiasts, to work at it.

The interaction of work and fun was illustrated in the 1874 *St. Nicholas Magazine* story "The Seaweed Album" by "Delta." In the story, Alice and her siblings convinced Mama to take them to the beach so that she could show them how to gather, preserve, and display seaweeds as her mother had long ago shown her. Once amidst the distractions of cool waves and warm sand, only Alice proved willing to work hard enough at the task to accomplish anything. Her siblings soon grew tired, and bored with play, but Alice enjoyed the day's work. Alice was rewarded for her work by having enough specimens to share with her siblings (who became interested when the work was done and they saw the results) and also to trade with other collectors. The message was clear: botanizing was neither all work nor all play, but rather a combination of both.[24]

The Legitimacy of Play

In the wake of the Civil War, Americans increasingly began to attach more value to play and fun. Shortened work weeks and child labor laws left members of the middle class with more time on their hands to fill. In response to this newfound freedom, sports and other hobbies boomed in popularity. Initially, interest in sports was tied to concerns about fitness and health, but before long the pleasure derived from them was considered equally important. Other hobbies and recreations followed suit and work began to take second seat to play.[25]

The new importance of fun and play, with which writers in the post–Civil War years increasingly struggled, is especially evident in Louisa May Alcott's *Under the Lilacs* (1878). In it two boys teamed up to botanize. The sickly but cultured Thorny, who was familiar with botany, read off characters of plants while keying them out. The healthy but unschooled Ben, who was supposed to record the characters and learn botany in the process, found the activity intolerably dull and tried to quit. The endeavor was too much like work, and the author had already made clear that Ben shied away from unnecessary work. Thorny sympathized, but argued that botanizing was more than just work: it was also fun when one got the hang of it. Indeed, Ben soon became an enthusiastic convert, emulating the youthful Wilson Flagg, tearing over the countryside collecting specimens for Thorny. Ben regaled Thorny with tales of his adventures. Thus, Thorny vicariously experienced the finds in much the same way Wilson Flagg's sisters had. Ben never shared the "capital fun" that Thorny found in studying, classifying, and book-bound botanizing. The pleasure Ben took in collecting was derived from his love of physical activity and his devotion to Thorny. Unlike Rollo, or the youthful Wilson Flagg, or even Thorny, Ben was deterred, not challenged, by toil. His love of collecting was love of play, not of work well done.

Alcott's Ben represented a new sort of child. His life included activities—from baseball to birthday parties—that Rollo never dreamed of. Ben was far less dependent upon adult authority than were Rollo and his peers, and he chose his recreations from a far larger pool than Rollo's. Ben worked only enough to earn his keep, with little concern that good work habits might ensure his salvation or make him a productive member of society. Ben was not lazy; he did, however, distinguish clearly between work and play, unlike Rollo whose play was work. In this way, Ben reflected changes in the social fabric that had occurred in the years between the creation of Rollo in the thirties and of Ben in the seventies. In those four decades Americans, especially the middle class for whom and about whom Jacob Abbott and Louisa May Alcott both wrote, had learned to take vacations, to play and watch sports, to ride bicycles, and to consider these and other leisure activities important.[26] Writers on botanizing had argued for most of the century that botany should not be reduced to drudgery by overemphasizing book learning and memorization. In the final decades of the century, authors took the notion that botanizing *could* be fun a step further by arguing that botany *ought* to be fun, and that pleasure alone was a legitimate reason to botanize.

The impact of the late-century belief that botanizing should be fun was far-reaching. One of the clearest examples of the increased emphasis on fun came in 1891, when Indiana school teachers read a botany textbook as part of a summer enrichment program. The book was an especially dry one, and the bulk of the teachers declared that they wanted no more of the science, not because they saw it as unimportant or useless, but because it was dull. Even as a school subject botany was, by the nineties, expected to be fun, pleasant, and interesting.[27] Writer after writer in the eighties and nineties waxed eloquent about the pleasures of botanizing with far greater force and frequency than their predecessors. Botanizing gave a pleasant purpose to other outdoor activity: walking, riding, or medical rounds. It deepened one's overall enjoyment of nature. One writer remarked that above all other reasons for botanizing reigned the "pleasure of knowing something of the life and activity going on around us."[28] These were not new claims; they were merely recast in the light of an increasing approval of leisure.

The playful aspect of botanizing was helpful to late-century teachers who worried about where to fit botany into tight school schedules. The popular times for botany lessons were the first thing in the morning and just before a recess. The logic was similar for both: proponents of the prerecess plan argued that students would be inspired to spend recess looking for specimens; proponents of the morning session argued that this enabled students to collect specimens on the way to school. "In his search," the child "will have the company of his classmates—there will be fun in it. They will exchange and talk them over before school. More is often learned thus than in the class, without thinking of its being study."[29] Both schedules relied on the enthusiasm of students to supplement work time by choosing botany over conventional forms of play, not as an extension of their work but as part of their play. The aim was not unlike that of Asa Gray's recommendation for extracurricular botanizing, but, unlike Gray's plan, it unabashedly capitalized on the appeal of play.[30]

The Work of Professionals, the Play of Amateurs

As botanizers adjusted their attitudes about work and play, their relationship with professional botanists changed. Professionals perceived botany as work because it was their full-time calling and often their employment. Amateurs, seeking only personal enrichment, increasingly saw botanizing as leisure, and deemphasized the worklike qualities. An 1897 editorial in the *Plant World* addressed the role of amateurs in a way that would have been inconceivable thirty years earlier: "The word amateur is often applied as a mark of reproach to those who are unskilled or mere dobblers [*sic*] in an art or science. More correctly it should be applied to one who pursues a subject from taste or attachment, without a view to gain; in other words from the mere love of it."[31] The notion that amateurs did botany for the love of it was not unique to the *Plant World* nor was it new, as the original meaning of the word, "one who loves," reflects. Remember, for example, Alphonso Wood's words: "Toiling for hours as no schoolmaster could have compelled us to do, being attracted to the task for the very love of it alone."[32] But what *was* new in the *Plant World* and other journals was the notion that amateurs of the eighties and nineties enjoyed botany as a recreation and not as a work surrogate. The *Plant World*, the *Botanical Gazette*, and the *Observer* all addressed amateurs in their pages with a consistent message: amateurs should botanize only if they enjoyed it. They should not aim toward contributing to science. They should not subject themselves to boring laboratory work. They should focus on field botany and observation. They need not spend all of their free time botanizing, but realistically might choose botanizing to fill a "few hours of an occasional holiday." Botany was to be a hobby, something one dabbled at, because it was fun. In an article on "Botany's Charms," the botanical writer of the *Observer* advised in 1891 that botanizers need not spend all of their time on analyzing plants and cataloguing their work, but rather should study what interested them, whether it was plant uses or nomenclature.[33]

In part, the message from the journals that amateur botany should be fun was an adjustment to professionalization. As professionals' focus shifted to the New Botany, they became less dependent upon the fieldwork of amateurs. Amateurs could no longer look at their activity as a scaled-down version of the work of professionals. Amateurs were also growing ever less concerned about the worklike qualities of botanizing, and more interested in pleasure, relaxation, and other recreational attributes. The advocates of botanizing in mid-century had stressed the worklike rigor, discipline, and devotion required to make it productive. By late in the century, advocates instead stressed how much fun and how playlike it could be. Late-century amateurs and their advocates complained bitterly about dry texts and teachers, and echoed Emerson's sentiment that professionals "love not the flower they pluck, and know it not. All their botany is Latin names."[34]

The new acceptance of amateurs' seeking fun rather than work was exemplified in a modest book, *How to Know the Wild Flowers* (1893). Its author, Mrs. William Star Dana, was a New York socialite and nature lover. The work sold 43,000 copies by 1899, and is still in print today. A quotation from nature writer John Burroughs at the beginning of the book spelled out its mission: "One of these days some one will give us a hand-book of our wild flowers, by the aid of which we shall be able to name those we gather in our walks without the trouble of analyzing them."[35] Dana's book eliminated "the trouble" for the common wildflowers of the Northeast. The author briefly described common flowering plants, which she grouped by petal color. Many descriptions were accompanied by an illustration that merged botanical accuracy with art; others included a few paragraphs of prose and poetry on the plant's habits, its beauty, its use, how it got its common name, or other brief facts of general interest. The book was remarkable in its lack of scientific trappings, offering just enough to distinguish the plant (and sometimes not quite enough to do that). Its popularity lay just there, in its simple but relatively accurate presentation. *How to Know the Wild Flowers* allowed the botanizer ready access to botanical information, presented in an easy-to-use format understandable to the most nascent beginner. Dana's accomplishment was to remove most of the work from one branch of botany. Nowhere did she suggest preserving specimens for reference, and rather than systematic record keeping she recommended keeping a notebook with random jottings for stirring the memory.

Dana's *How to Know the Wild Flowers* was clearly designed for the amateur. It became popular because it unambiguously portrayed botanizing as a recreation. Oakes Ames, Harvard botanist from 1898 to 1941, recalled Dana's book as having been influential in luring him to take up botanizing as a youth, despite the fact that he was somewhat lazy and not academically inclined.[36] Behind Dana's book was a social environment that demanded that botanizing be above all interesting and fun, whether in the classroom or as a hobby. While Wilson Flagg and Alphonso Wood enjoyed collecting in part because of the challenge and hard work involved, young Oakes Ames pretended he was exploring strange, distant, make-believe lands. Botanizing competed for his time with football and baseball, rather than with chores or schoolwork. Only slowly did botanizing win out. The worklike qualities of botanizing had long contributed to its popularity; now its recreational value served to attract converts.

The attitudes of Americans, especially of the middle class, toward work had undergone dramatic changes between the boy-hoods of Wilson Flagg and Oakes Ames. There was a diminished concern with filling all time with work or its equivalent, and a growing recognition that both children and adults needed play. Botanizing could be work, play, or both. No small part of the continued popularity of botanizing lay in this versatility. Its successful transition from work surrogate to recreation illustrated this. During the years of transition, the pleasant features and worklike qualities of botanizing combined to make it a desirable pastime.

As amateurs began to consider their botanizing less like work, the line dividing amateurs and professionals became firmer. The same editorials and articles in journals that urged amateurs to shape their activities to what was fun, urged professionals to shape theirs by what they considered important—boring and tedious as that might be. This was only one of many growing differences between amateurs and professionals. It was, however, an important one, for it partially explained the gradual devaluation of amateurs in the scientific community. No mid-nineteenth-century professional could have survived without specimens and observations provided by amateurs. In contrast, in the early twentieth century, few professionals relied on amateurs. This was not because there were no longer any amateurs, but because the activities of amateurs were no longer modeled after those of professionals. The portrayal of botanizing as play, whether or not it actually affected the type or caliber of botanizing done, gave the appearance of a less serious and less scientific approach to the science at a time when scientists were extremely concerned about standards. Indeed, late-century amateurs who did send specimens to professionals often went to great lengths in describing the difficulties encountered in securing specimens, as when Kate Furbish wrote to

Asa Gray: "I climbed Mt. Blue on Friday, and found the shoots which I send you. It was 12 hours before I could put them in water, and the leaves were so tender that they were sadly wilted but now they are fresh again, and I trust will not be so wilted that you cannot see them in their beauty."[37] Despite the exceptions, like Kate Furbish, the increasing division between amateurs and professionals over attitudes about work and play can only have contributed to the already growing gap between the two groups.

Thus, attitudes about work and play were a major influence on the popularity of botanizing and also on the sort of botanizing done. The evolution of botanizing from work surrogate to recreation kept it popular during an era of transition, but its status as a recreation had mixed results. On the one hand, it allowed botanizing to compete with other leisure activities for votaries. On the other hand, it contributed to a decline in the status of amateur botanizing within the professional botanical community, boding ill for the health of amateurism.

7
Natural Theology and Amateur Botany

The botanical community, both professional and amateur, of the first half of the nineteenth century uniformly believed that the pursuit of botany inevitably led to a better understanding of God. Doubts, if there were any, remained entirely unexpressed. Natural theology, the idea that the existence and character of the Creator are revealed in the natural world, pervaded American botanical thought. For amateurs, the belief that coming to know the Creator's works would lead to greater piety made botanizing one of the most compelling means of self-improvement. Moreover, in an era heavily influenced by Protestant theology, sciences like botany thrived because they posed little threat to the harmony of science and religion. Finally, early- and mid-century Protestant scientists and philosophers favored an empirical approach, which was compatible with the heavily taxonomic focus of American botany. Botanical natural theology was, in short, ideally suited to the religious, scientific, and cultural moods of antebellum America.

As the century progressed, however, both the scientific and the theological frameworks underlying natural theology changed. Once a central theme of botanical writing, natural theology became more and more tangential. As professionals abandoned an open adherence to natural theology, they increasingly viewed it as a distinguishing characteristic of amateurism. Simultaneously, the appeal of natural theology as a form of self-improvement diminished. By century's end, even amateurs were losing interest in looking "from nature to nature's God" as part of their botanical pursuits.[1]

Antebellum Piety

The popularity of natural theology in early- and mid-nineteenth-century America is readily illustrated by its prominent place in education. From common schools to colleges, students received instruction in natural theology across the curriculum. Schoolbooks, especially readers for younger students, liberally sprinkled their pages with natural theology to remind children that behind nature there stood a wise and beneficent Creator.[2] Pedagogues, always on the watch to improve their charges, used poems, parables, and essays to convey such messages as "though the rose is beautiful, He who made it is more so."[3] Topics like "The Goodness, Wisdom, and Power of God" were common fare, instructing that "if every living thing, and every shrub, and flower, had written upon it that God is good, and wise, and powerful, His goodness, wisdom and power, would not be a tittle plainer to be seen than they are now."[4] Even fiction illustrated the wisdom of God's design of the universe, as in one piece in which an acorn drops painfully on the head of a man who has just wondered aloud why big fruit does not grow on trees.[5] Few nineteenth-century school readers were untouched by natural theology, and few students could have failed to encounter it at a young age.

This early introduction prepared students well for botanical texts, which frequently drew attention to natural theology. Early in the century, Amos Eaton's *Manual of Botany* claimed that "the mind is naturally led from the contemplation of the beauties of creation, to that intelligence and power which gave them birth."[6] Almira Lincoln Phelps was even more direct in her widely used *Familiar Lectures on Botany*: "The study of Botany," she wrote, "naturally leads to greater love and reverence for the Deity."[7] No botanical author, however, rivaled Asa Gray in his devotion to presenting natural theology to students. In the introduction to *Botany for Young People*, Gray stressed the special relationship he saw between botany and religion, a central theme in each of his texts: "This book is intended to teach Young People how to begin to read, with pleasure and advantage, one large and easy chapter in the open Book of Nature; namely that in which the wisdom and goodness of the Creator are plainly written in the *Vegetable Kingdom*."[8] For much of the century the connections of botany with natural theology contributed heavily to the popularity of the science as a school subject.

Texts on natural theology were also standard reading at many colleges, and in some academies. William Paley's *Natural Theology*, which went through more than fifty American printings between 1802 and 1865, was among the most popular textbooks on any subject for academy and college students in antebellum America.[9] The inclusion of Paley among the textbooks that a school used indicated both orthodoxy and high standards well into the century, and his basic ideas formed the core of American natural theology.[10]

Through his writings and their popularizations, William Paley (1743–1805), a British Anglican clergyman, introduced several generations of Americans (as well as Britons and Canadians) to an Enlightenment-tinted version of natural theology. Seen through Paley's eyes, God was a kind and wise provider. Logic and reason allowed human observers to deduce the existence and nature of the Creator using evidence found in the "Appearances of Nature." Central to Paley's work was the "argument from design," which posited that just as an observer finding a watch infers a watchmaker, so a naturalist studying an eye or a hand infers a designer—namely, God. Paley further argued that one could deduce from the variety, beauty, and usefulness to man of the natural world that God was wise, powerful, and beneficent.[11]

Antebellum botanizers adopted Paley's argument from design and used it in tandem with another old and established tenet of natural theology, the doctrine of "divine providence," which ar-

"And God Saw That It Was Good" (Frontispiece from Almira Hart Lincoln Phelps, *Botany for Beginners*, 1891)

gued that God had not only created the world, but also continued to sustain and control it. For the naturalist, "special providences" included such miraculous events as God's direct creation of the earth or of species. "General providence" proved a useful tool in explaining God's work through the agency of natural law, such as the purification of air by plants. [12]

Natural theology found a solid base in the dominantly Protestant culture of antebellum America. Almost all American intellectuals, certainly most of the scientific leadership, were orthodox

Protestants, which entailed, among other things, two beliefs especially conducive to natural theology: first, God had documented the Creation of the natural world with obvious evidence in nature; second, God had provided a reliable, though perhaps symbolic, record of the Creation in the Bible. Because the same author wrote both accounts, natural and revealed theology must be in harmony; science was therefore an ally, not an enemy, of religion. Natural theologians attempted to unify knowledge and belief in the hope of producing a religion free from doubt, a "scientific" theology.[13]

Similarly, natural theology was at home with empiricism, the dominant trend in early-nineteenth-century scientific thought. The scientific community embraced Francis Bacon's call for empirical, antitheoretical, and—most significantly for botanic studies—taxonomic work.[14] Baconianism appealed to the broad spectrum of American Protestantism, because it involved the facts of God's creations rather than mere speculation. More importantly, it fostered precisely the kind of science that most amateur botanizers pursued—namely, empirical observation and classification. Thus, it is hardly surprising that many antebellum American botanizers subscribed to natural theology.

The appeal of natural theology as a form of self-improvement was also deep and widespread. Self-improvement was a vital force behind botanizing, and for antebellum botanizers, no other form of self-improvement—not gentility, or utility, and certainly not mental discipline or physical culture—appealed more than the promise of greater piety and stronger faith. The strongest theme of reform argued that spiritual well-being and revival are essential to creating moral individuals and a moral society. This gave botanizers a special devotion to natural theology: while gentility and utility promised much, natural theology promised far more.[15]

The spiritual, scientific, and reform climates of this period each contributed to the infusion of natural theology into all areas of science. Yet, despite the popularity of natural theology in other disciplines, botanical writers felt that their science was the best suited to illustrating God's wisdom and character. Just what shape this special relationship between natural theology and botany took, and how it reflected changes in both the scientific and religious cultures of the nineteenth century, can be seen in the content of botanical natural theology. Botanical writers addressing natural theology uniformly began with the assumption that God wanted people to study *both* the Bible and nature. [16] Unlike geology or astronomy, pre-Darwinian botany had the advantage of having no apparent area of contradiction with the Bible, a problem that consumed a great deal of energy in some disciplines. [17] So harmonious was the relationship of botany and the Bible that several botanical texts began by recounting the Mosaic account of the creation of plants. The frontispiece of Almira Lincoln Phelps's *Botany for Beginners* depicted the introduction of plants on the third day of Creation with the caption " 'And God saw that it was good'—Gen. I. 12." [18] Only a little less bold was Asa Gray's statement on the harmony of revealed and natural theology in his *Botany for Young People*, which began with Matthew 6:28: "Consider the lilies of the field, how they grow." [19] Lest any reader miss the point, Gray continued: "Our Lord's direct object in this lesson of the Lilies was to convince the people of God's care for them. . . . And when Christ himself directs us to consider with attention the plants around us,—to notice how they grow,—how varied, how numerous, and how elegant they are, and with what exquisite skill they are fashioned or adorned,—we shall surely find it profitable and pleasant to learn the lessons which they teach." [20]

The assertion that God wanted people to study botany was not uncommon and was not limited to authors as successful as Lincoln Phelps and Gray. Botanical authors of all ilks argued that God had intentionally designed plants to reveal His ways. It was not, as "A Lady" wrote, simply chance that caused botany to reveal God's ways, but rather part of God's design: "Looking down I found delicate wild flowers, which . . . seemed to whisper a lesson of love for Him, who has manifested His goodness, by placing forms of beauty everywhere that man might chance to wander; forms that would woo him to the study of glorious nature, and from this to a study of nature's great Creator." [21]

To the antebellum botanizer, finding evidence of design in the plant world almost required going on to look for the Designer. Noting that some readers failed to observe the connection, many authors cautioned that while God had written the book of nature, people still had to interpret it. Despite her conviction that botany led to "love and reverence" for God, Almira Lincoln Phelps, with her reformer's compulsion to help her readers improve themselves, spoke bluntly about the necessity of actively looking for God's hand in the natural world: "there are some minds which, though quick to perceive the beauties of nature, seem blindly to overlook Him who spread them forth. They can admire the gifts, while they forget the giver. But those who feel in their hearts a love for God, and who see in the natural world the workings of His power, can look abroad, and adopting the language of a Christian poet, exclaim, 'My Father made them all.' "[22]

Far less eminent authors than Lincoln Phelps also stressed the need to look for the designer behind nature.[23] Even a schoolbook essay on George Washington could be used to illustrate the importance of natural theology, through a story in which Washington's father spelled out his young son's name in seeds to show George that design is evidence of a designer. George, who obligingly drew the connection between his father's role in the garden and God's role in the universe, was allegedly left with a lifelong conviction of "God the author and proprietor of all things."[24]

Behind these urgings to study God's creations in order to understand God better lay the certainty that the inevitable result would be moral revival in both individuals and their society. Botanical natural theology, its supporters claimed, eliminated misanthropy, peevishness, and malice from its followers.[25] Educators, in particular, hoped that teaching children to read God's book of nature would result in their learning to love its author. Thus enlightened, children would be able to ward off evil and vice, would be free to exercise benevolence and mercy, and would thus mature into "more virtuous and happy citizens."[26] This spiritual enlightenment, and the resulting improvement in morality, appealed highly to pious, reform-minded, antebellum Americans.

Botanizers used the argument from design, first, to document the existence of God, and second, to show God's character. God's authorship seemed to be plainly written throughout the plant world, as "His pencil glows in every flower."[27] One author put it simply in an article for children, stating that the study of flowers "will show us the hand of God in their production."[28] Another painted a far more dramatic picture, telling of an atheist who, while studying a rose, suddenly exclaimed, "What a fool I am! Here I am trying to make myself believe there is not God, when I see the mark of His fingers on everything around me." The new believer "became a most devoted Christian. He burned up his infidel books, and became warmly attached to the Bible."[29] Varying more in the way they expressed themselves than in their intended messages, a number of authors proclaimed to their readers that after studying botany "none but the fool can say 'There is no God.' "[30] This was a very pronounced form of self-improvement indeed.

Many authors, assuming that the reader accepted God's authorship of nature, emphasized what design in the plant world revealed about the character of God. One botanist wrote: "On no page of creation, can be found more distinctly written, the wisdom, benevolence, and love of the Creator, than on that which exhibits the structure and adaptation of organization to circumstances, of the humblest vegetable."[31] The provision of material for clothing, food, and shelter also served to demonstrate the wisdom and goodness of the Creator.[32] The evidence of God's nature was so strongly revealed in each leaf and flower that, "if there was nothing else to prove the wisdom and goodness of God, these would do it amply."[33] Who but a kind and wise Creator could have provided for all human needs? And God did not stop at supplying necessities, but added adornment—beautiful flowers, for example—to give pleasure. This mingling of utility and beauty struck nineteenth-century botanical writers as one of God's greatest gifts, revealing plainly a "tender goodness" and divine love.[34]

Although botanical writers used the argument from design with great frequency and fluency, they were less comfortable and less successful in demonstrating providence. Design could be shown by single examples, taken from contemporary, living specimens, but providence was most commonly illustrated by change. American botanizers focused on empirical taxonomy, rather than on aspects of botany that lent themselves to demonstrating providence, such as plant growth or the creation of plants. One area where botanists did use providence was in their explanation of the adaptation of plants to their environments. As one author argued, the adaptability of plants was evidence of God's continual orchestration of the world: "Changing plants from one climate to another greatly modifies them, so much so that new species have apparently been produced by artificial means. In nothing more than this power of the vegetable to adapt itself to various circumstances, is the wisdom of the Almighty seen."[35]

Like their contemporaries in astronomy, botanizers used general providence to include action through natural law as well as the traditional use of special providence through direct action or special creation.[36] In one children's story, a girl wonders whether

God really makes each flower. In answer she learns that God is as much the maker of flowers as a machine operator is the maker of the products the machine turns out: "And so God makes the flowers by means of the machinery of his laws, or what are commonly called the laws of nature."[37]

Botanical writers also demonstrated providence by citing the ability of plants to transform their environment and thus to make the earth habitable and pleasant. Describing the role of green plants in improving air, one author reminded readers to be thankful for "the wisdom of the system."[38] Another queried rhetorically: "Who teaches the flower to bloom?"[39] The role of Spanish moss in absorbing moisture from the air, "and thereby rendering Louisiana 'tolerable,'" struck one author as a sure sign of God's providence.[40] Others who lived perhaps in more pleasant climates were more apt to comment on God's powerful use of plants to purify and add oxygen to the air, or to transform a "barely cool, rocky planet, into a home for man."[41] These ongoing changes in and by plants seemed ample evidence not only of the wisdom of God's design, but moreover of the continued actions and governance of a powerful God.

Religion vs. Science in the Later Period

Despite its broad base and widespread acceptance, natural theology encountered changes in both theological and scientific beliefs as the century progressed. These changes affected the professional community more directly and quickly than the amateur community, but their effect on the amateur community was nonetheless profound. And although the difficulty of reconciling religion and science occurred largely in other areas of science, the changes wrought in other disciplines eventually affected botany. Debates over the age and history of the earth and the universe, for example, so weakened the forceful focus of natural theology on causation that by the 1850s it was giving way to watered-down documentation of the harmony of science and religion. In botanical circles, this weakening was reflected in an increased use of natural law to show providence. At first, this took the form of an increased use of general providence, but gradually botanizers placed less emphasis on providence and more on the laws themselves. The broad support from religious education turned elsewhere, and the strong movement that had been natural theology was left weak and diffuse.[42]

These scientific and religious challenges left botanical natural theology ill equipped to survive the mid-century changes in ideas about self-improvement, which gave less emphasis to religiosity. Those who sought moral reform, whether from the Enlightenment's confidence in progress or from the Romantic belief in perfection, were far less vocal than they had been. Those who sought personal advancement in the new mold (and who formed a large core of botanizers) were more attracted to other attributes—for example, physical culture and mental training. From the perspective of theologians, natural theology was no longer quite so certain to lead on straight from the Creation to the Creator.[43] From the perspective of professional botanists—with important exceptions, such as Asa Gray—natural theology was no longer part of science and hence ceased to be a means of promoting science as a tool for self-improvement.

The scientific challenge, professionalization, led to a public disdain by professionals for the questions and methods of natural theology. At the core of professionalization was the identification of and adherence to a specific body of knowledge and special methods used to create and study it. Unlike their forebears, who had seen science as an ally of religion, late-nineteenth-century professionals believed that science and religion belonged to separate intellectual spheres. Hence professionalism increasingly came to mean, among other things, that science and scientists avoided commenting upon religion, at least in scientific works.[44]

Yet for amateurs at mid-century, natural theology remained a widespread interest. The authors who espoused it ranged from botanical novices to experts, came from all walks of life and all regions, and showed little awareness of the revolution in the relationship between professional science and religion. This can be at least partially explained by the continuity of one of the earlier sources of support: the leadership of Almira Lincoln Phelps and Asa Gray, whose works remained popular long after their deaths, until century's end. Gray's role is especially interesting, because he not only was the most articulate advocate, but also was among the last of the professional scientists who openly advocated natural theology in their scientific publications. His accommodation of

Darwinian evolution to natural theology may have contributed to the ability of amateurs to maintain their belief in an era when natural theology was in decline. Amateurs' continued interest in Gray's textbooks, all of which endorsed natural theology, kept design in the public eye in botany long after it had disappeared from other disciplines.

Gray's contribution was to expand the use of natural law from providence to design. He argued forcefully, as had others, that it was entirely plausible for God to have used natural law beneficently designed and guided to create the world and its inhabitants, rather than having created them directly. Thus, observers ought to look for evidence of design and providence in law as well as in fact, in mode of formation as well as in individual structure. Further, Gray argued that this accommodation amply allowed for a changing world. He expanded the role of natural law and, equally importantly, he delivered his message clearly to both the public and the scientific community through a series of essays that were published in popular as well as professional journals, and were later collected in book form. [45]

Just how much of Gray's subtle argument amateurs comprehended is unclear. What is clear is that America's foremost botanist, the chief supporter of Darwin and a leader of the professional scientific community, was unflaggingly committed to natural theology. Like their mentor, American amateurs remained faithful to natural theology and botanizing. Many botanizers did adopt Gray's general line of reasoning that organic change, natural selection, and especially adaptation were part of the Creator's design and providence. One author argued that the symbiotic relationship between insects and orchids was evidence of "all wise design."[46] More forcefully, one fancier of parasitic plants argued: "the adaptability of living forms, even the simplest, to their uses, surroundings, and needs bears powerful testimony to the existence of an Infinite Mind, and the highest value of Natural History studies is that, if read aright, they teach us to look from Nature up to Nature's God."[47]

In the last decades of the century, while some botanizers turned their attention to adaptation, the more common response was an erosion of interest in natural theology. Lacking the endorsement of prominent religious leaders and of most professional botanists, natural theology lost much of its force. Late-century amateurs offered only weak statements of endorsement—or, more often, no mention at all. While antebellum botanizers had emphasized design and providence, their successors were more apt to argue that studying nature teaches about nature's God, with little or no discussion of how or to what end the Creator worked. To William L. Baily, writing in 1870, it seemed self-evident that God had designed plants "for our enjoyment" and that not to study them would be disrespectful. Yet, Baily's ardent hope was that the reader's mind would be turned from "the bright world around it, to the great Fountain and Source and Creator of all," from "the visible objects of its admiration, to the adoration of the Invisible who created them."[48] Statements like this had been common in the early decades of the century, but they were exceptions in scientific literature by the seventies. Moreover, even the strongest endorsements of natural theology of the late nineteenth century paled in contrast to the ardent endorsements of earlier decades.

More typical of the late-century attitude was the practice of prefacing or concluding an essay with a superficial statement of support for natural theology. Replacing concrete demonstrations of design and providence were token endorsements that "the creations of this earth are intended to become revelations of the greatness of Omnipotence," or that "nothing shows more plainly the tender goodness of our Creator."[49] Detailed accounts of natural theology became increasingly rare, although it was not yet unusual or unacceptable to express hope that a botanizer might be led to see "the goodness and glory of the Creator" or to look "from nature up to nature's God."[50] Just as early-nineteenth-century writers often gave token endorsements of utility for the sake of public appeal, so did later writers swear token allegiance to natural theology. Certainly the fervor of natural theology had died. The professional community by and large ceased to discuss natural theology in scientific circles when Asa Gray died. While some professional botanists may have privately adhered to natural theology, professionalization entailed a public separation of science and religion. Even Asa Gray was reluctant to draw scientific meaning from the Bible and thereby to judge science by its conformity with revealed theology. But while professionals found natural theology objectionable in their own ranks, they ignored its continued popularity among amateurs.

While natural theology still had symbolic value, by the end of the century it lacked the appeal it had once held for amateurs. Self-improvement had lost much of its religious compulsion. Botanical leaders like Asa Gray no longer reminded readers that God charged them to study botany, nor promised that botanizers would understand their Creator better by studying His creations. There were other reasons to botanize, but none with the universality and compulsion that natural theology had possessed. With the demise of natural theology, amateur botany lost a valuable ally.

8
Botany and the Rhetoric of Utility

Hand in hand with antebellum enthusiasm for other varieties of self-improvement went a strong interest in self-sufficiency that encouraged utilitarianism. The acquisition of practical knowledge, including applications of botany, was one means of becoming a less-dependent, and therefore a better, citizen. The most frequent claim to the utility of botany was that the study of plants is inherently practical, since humans use plants for food, shelter, clothing, and medicine. Texts, reviews, and magazine articles were full of general statements like "Our food, our medicine, our luxuries are improved by it," or "A vast number of plants are of great use to man and beast," or "Every thing in the kingdom of nature has its use, and to know something about the utility, conduces to render us more useful."[1] The strength of the appeal of utility is clear from the prevalence of these statements.

The promoters of botany targeted two areas as particularly

linked to the practical side of botany: agriculture and medicine. General statements about the connections of these two areas to botany abound. One Southern observer commented: "The planter would find important links as to the cultivation of his fields and garden—the physician would find new simples, safer and more efficacious than adulterated foreign drugs."[2] One writer for girls, describing the qualifications for marriage and motherhood, observed that a knowledge of "botany will prove of great use to you, with respect to an acquaintance with medicine. Every woman ought to understand the elements and composition of the remedies provided for her, or by her administered to others."[3] Clearly, such links did exist in a day when the nation was predominantly agricultural, and when medicine included widespread domestic and professional use of botanic drugs.

Agriculture

For those interested in agriculture, the hope was that a knowledge of botany would promote better farming practices. For example, John Darby's *Botany of the Southern States* opened with a ringing endorsement of the importance of botany: "To an agricultural people, there can be no subject more important, or really demanding a deeper interest than Botany. . . . It indicates the conditions essential to the growth and perfect development of plants, their food, the means of supplying it, the condition in which it must be furnished, and the methods best calculated to gain a given result."[4]

Specific agricultural applications illustrating the usefulness of botany to agriculture included the physiological aspects highlighted by Darby, but the focus was more often on more basic issues, such as crop selection. Observers noted with interest the great expansion in the variety of plants being grown commercially and pointed out that correctly matching a given variety to a particular locale was of great importance. Understanding the effects of nutrition, climate, and soil on the plant was crucial. Fruit growers and vintners were particularly sensitive to the issue, perhaps because mistakes affected them for more than one growing season.[5] Others pointed to the importance of a knowledge of botany in the introduction of the potato to France, which "in all probability twice saved France from famine."[6] Great concern existed about timber crops and ornamental trees. One oft-cited instance where a lack of botanical knowledge had led to unhappy results was the case of the Lombardy poplar: those who introduced the tree to America had brought only the staminate trees, not realizing that there were male and female plants; the tree can be propagated by root cuttings, but each succeeding generation produced this way is less and less desirable, whereas those produced from seed breed true.[7] Writers on timber production stressed the importance of using the right sort of wood for the right purpose, and the necessity of being able to recognize the living tree and the milled timber.[8]

These concerns with crop selection and adaptability carried over to both ornamental gardening and vegetable crops. Topics such as the possibility of cultivating wild orchids, based on a knowledge of their needs, and the use of an understanding of botany in selecting varieties of home crops, were frequently discussed in agricultural and popular journals.[9] Cautions about the necessity of understanding the life cycle, and of being able to recognize the seed and the young plant of a particular crop, were also common. William Darlington, a frequent commentator on agricultural topics, illustrated his warning that a farmer with no knowledge of botany was susceptible to the influence of many myths that would limit his success, with the example that a farmer who believes in the transmutation of species will accept seed that is not "clean," or has some seeds of another species mixed in.[10]

Similarly, the usefulness of botanical knowledge as an aid to weed identification received much attention. William Darlington drove home the value of distinguishing species of weeds in a parable in which a farmer expended a great deal of effort to attack a relatively innocent weed, while an especially noxious one thrived nearby with disastrous results.[11] Another author, more prescriptively, recounted a conversation between himself and a friend: the two realize that neither of them knows the name of the common weeds they see; the friend resolves to spend the summer teaching himself plant identification, because he wants to know the names of the visitors on his property.[12] Agricultural journals, especially the *Country Gentleman*, often beseeched their readers to pay closer attention to botanical matters by stressing, as had Darlington, the benefits of prompt identification of weeds.[13]

Yet another practical side of botany was the ability to identify

"The Farmer and the Class in Botany" (From *Youth's Cabinet* 4 [1849]: 283; courtesy of the American Antiquarian Society)

poisonous plants, especially those that resemble edible ones. Children's books were filled with warnings about the importance of being able to identify poison ivy, but also with the importance of knowing what you are eating. In one cautionary tale, children were told of "a poor boy, who in rambling about the fields with his little brothers and sisters, chanced to meet with a root of a hemlock-dropwort. It looked so white and nice, that he was tempted to eat a good deal of it. The other children also ate some, but not so much. When they got home they were all taken very ill. The eldest boy, who had eaten most, died in very great agony. The others recovered, after suffering a great deal."[14] Special cautions were aimed at those, young and old, who sought to collect and consume mushrooms. Authors stressed that the only safe way to distinguish accurately between poisonous and edible mushrooms was to identify them botanically, discarding old wives' tales and mastering, "one by one, the specific distinctions."[15] Similar discussion of the uses of lichens (including their use as food) also received attention—though, understandably, not as much.[16]

Amidst the enthusiastic endorsement of the utility of botany, however, there were skeptics—like the writer for the *Country Gentleman* who expressed a caution: "We do not think the study of botany will enable the farmer to become 'rich' without the best practical knowledge, nor generally even to raise larger crops."[17] And John Darby's confidence in his claim that "there can be no subject more important, or really demanding a deeper interest than botany," would be more believable if his text provided the practical information it promised rather than mere descriptions of native species. Indeed, the utilitarian appeal of agricultural botany lay more in promise than in practice. Nowhere is this more clearly illustrated than in the relative lack of botany in the agricultural curriculum as the land grant colleges emerged. Long before the advent of agricultural education, William Darlington had advocated botanical education: "If there be those who do not wish to inform *themselves* beyond the manual operations of the field and barnyard, it is at least due to the future good standing of their *children* in an intelligent community, that the youths should have some chance of escape from the chrysalis condition of darkness and prejudice."[18] The *Country Gentleman* would take up the same cry in calling for a "Liberal Education for Farmers": "It does not signify, because those who were before us knew nothing of chemistry, of geology, of botany, of mineralogy, of meteorology, and the improvements in steam and machinery, that we, their descendents, should know nothing of them too."[19] Unfortunately, when botany entered the agricultural curriculum it appeared as horticulture and plant pathology; pure botany entered only later, as individual botanists like Charles Edwin Bessey found themselves teaching at the land grant schools.[20]

Medicine

Just as one would have expected to see links between popular botany and agriculture, one would expect the rhetoric about the connections between botany and medicine to translate into reality, particularly during the antebellum period when botanic medicine was an important thread of both professional and domestic medicine. Yet if few antebellum Americans benefited directly from the agricultural applications of botany, medical links were even more tenuous. Despite the importance of plants in folk and domestic medicine, neither the literature of popular botany nor that of domestic medicine provided tightly drawn connections between the two disciplines. Historical questions that begin "why didn't . . ."

are impossible to answer in any conclusive way, but examining the context does provide some clues.

Domestic medical manuals rarely gave readers enough botanical information to ensure proper identification. Jacob Bigelow's *American Medical Botany* and a few other major, expensive volumes did provide both accurate botanical descriptions and medical information, but they were the exception rather than the rule. Surprisingly, even the literature of Thomsonianism, the popular botanic domestic sect, was largely devoid of botany, focusing instead on therapeutics; for example, even the description of lobelia in Samuel Thomson's own *New Guide to Health or Botanic Family Physician* would not have been sufficient to ensure that the collector had the right plant when collecting this most important of remedies:

> *The Emetic Herb* may be found in the final stages of its growth at all times throughout the summer, from the bigness of a six cent piece to that of a dollar and larger, lying flat on the ground, in a round form like a rose pressed flat, in order to bear the weight of snow which lays on it during the winter, and it is subject to be yellow and pale, like others suffering from wet and cold, but when the returning sun spreads forth its enlivening rays upon it, lifts up its leaves and shoots forth a stalk to the hight [*sic*] of from twelve to fifteen inches, with a number of branches, carrying up its leaves with its growth. In July it puts forth small pointed pale blue blossoms, which is followed by small pods about the size of a white bean, containing numerous very small seeds. This pod is an exact resemblance of the human stomache, having an inlet it receives nourishment and by the outlet it discharges the seeds. It comes to maturity about the first of September, when the leaves and pods turn a little yellow; This is the best time to gather it.[21]

This level of vagueness was by no means unusual. The entire description of bayberry in one manual read, "It is a kind of shrub; and generally grows from two to four feet high."[22] Other manuals dispensed with descriptions of plants altogether.[23] One noteworthy exception to the poor botanical information in botanic medical guides was Wooster Beach's *American Practice of Medicine*, the mainstay of eclectic medicine (the largely botanic practice that evolved out of Thomsonianism in the 1830s and 1840s). "Part VII: Materia Medica" consists of poor illustrations coupled with excellent botanical descriptions, including the Linnaean class and order, and descriptions of the genus and species that are detailed enough to be useful. The genus description for lobelia, for example, reads: "Corol irregular, 5-cleft, slit longitudinally on its upperside. Anthers united into a tube, Stigma capitate. Capsule inferior; 2 or 3 celled."[24] This level of detail, however useful, meant that Beach's work came in three volumes, making it expensive and cumbersome. The standard level of description suggests that Thomsonians and others interested in botanic healing either learned plant identification from other sources, or bought prepared remedies.[25]

The literature of botany, in turn, generally made at best passing references to medical applications. Just as agricultural applications were more often treated in generalizations than in detail, medical applications of botany appeared in superficial forms, like: "A knowledge of botany will enable you to ascertain, with certainty, the identity of plants, which are important as medical agents."[26] Indeed, some botanically oriented authors expressed mixed feelings about just who should pursue medical botany and how deeply. After stressing the importance of medical botany for practitioners, one author cautioned against its widespread popular pursuit on the grounds that it was too complex to be properly or practically pursued by lay people.[27]

Another author, who wondered whether, "on the whole, the knowledge of their [plants'] medicinal value is of much value," demonstrated the perils of a little knowledge by relating an encounter with a woman who assumed he was gathering herbs for medicinal use, there being no other reason to collect plants that she could understand. Despite his denials, she began taking specimens of everything he collected, with the intention of using them therapeutically without any knowledge of their botanical identity or their efficacy. As the author put it: "The method many of our good old domestic nurses adopt, when any of their children are sick, is to try the first herb or root that comes to mind, and if that does not effect a cure, to try another. . . . The old lady who took me for a *yarb* doctor, wanted to know if the plants I had were 'good for sick folks.' If they were, I suppose that was enough—and the poor victim under her treatment had to be dosed with them, no matter whether he had a cold or the small pox."[28] Even Almira Lincoln Phelps, who surely could have provided accurate botanical information on medicinal plants and who advocated using domestic medicine, ignored applications in her descriptions of plants. Instead, she promoted botany as a sure check against medical quackery, because it enabled patients to be more discriminating in selecting a practitioner—suggesting that physicians, at least, could be assumed to be botanically knowledgeable.[29]

And indeed, medical practitioners often did pursue botany. The ability to identify plants correctly in order to compound remedies was a skill of inestimable importance. Because of the central role of plants in the nineteenth-century pharmacopoeia, medical schools taught far more botany than did colleges, and well into the century they served as the educational training grounds of many professional botanists such as Asa Gray. Prior to the Civil War, separate courses in both botany and materia medica were often included in medical school curricula.[30] The connections between the two studies seem to have been drawn irregularly, at best. Only at medical schools established by eclectic physicians, who made heavy use of botanic remedies, does the subject seem to have been taught as "medical botany."[31]

Despite this apparent lack of commitment, physicians and the public evidently felt that botany was important to the practice of medicine. As one author put it: "To the physician it [botany] is nearly indispensable, since it makes him intimately acquainted with the external characters and medicinal properties of those plants reputed to possess healing powers."[32] The ability to obtain purer medicines seemed pressing for some, as did the hope of discovering new plant-based remedies.[33] The need to know existing materia medica encouraged some physicians to continue to botanize in a semiprofessional capacity after their school days.[34] Writers also expounded nonmedical reasons for physicians to botanize, dwelling on the honing of observational and reasoning skills that botany encouraged.[35] The nonmedical aspects of medical botany were exemplified by the suggestion that botanizing might help physicians by pleasurably filling the time spent driving on rounds, keeping their observational skills sharp, and preventing boredom.[36]

Other Applications

While medicine and agriculture were the most obviously practical—if not always utilized—applications of botany, promoters also linked botany with a variety of less plausibly practical activities. The spirit prevalent in nineteenth-century America that one should "be usefully employed, never be idle" led to many more esoteric claims for the utility of botany.[37] One author told a story of a shipwrecked traveler who washed up on some rocks: fearing that the tide would continue to rise and cover the rocks, he was about to reenter the water to look for higher ground, when he observed a plant that he knew grew only on land and reasoned that the rocks were safe.[38] Another suggested that a knowledge of botany was useful because it allowed one to distinguish between popular lecturers and writers who had something to offer, and those "who bepuff themselves in the newspaper" and "know scarcely the alphabet of the sciences which they undertake to teach."[39] Florists would benefit from knowing the geographical distribution of plants.[40] Others hailed the usefulness of a knowledge of botany as evidenced in its importance to the accurate painting of flowers and writing of travel guides.[41]

The generality of the many claims to the usefulness of botany and the rarity of botanizers who directly applied their hobby to practical uses suggest that while the appeal of utility may have been genuine, it was more rhetorical than real. Indeed, botany was not alone in its claims to being a utilitarian science. As professionals emerged in nineteenth-century science, they often noted, though not always favorably, that the public demanded practical science.[42] Two generations of historians of American science have debated the degree to which nineteenth-century scientists were or were not more interested in applied science than in basic research, without a clear resolution. This lack of clarity suggests that the scientists themselves were somewhat ambivalent. The strong rhetorical appeal of utility has sometimes been misread as description, rather than the prescription that it actually was. Historian George Daniels's suggestion that scientists used a number of claims, including utility, primarily in order to elicit public support and to ensure the social role of the scientist, seems to accurately represent, at least in part, the pattern (whether a conscious or an unconscious one) in botany.[43]

The Decline of Utility

Despite the infrequency of actual applications of botany in medicine and agriculture, the lengths to which authors went in demonstrating the utility of botany suggest real interest in utility on the part of some antebellum Americans. As the century closed, however, utility became less important as a reason to botanize. The interest in botanic domestic medicine declined, and botanical knowledge became less essential for the family head. Thomsonianism was no longer a popular domestic regimen. Two professional sects that emphasized botanical drugs, the eclectics and the physio-medicals, did practice in the last half of the century, but their members were even more apt to purchase ready-made remedies than their domestic forebears had been, and thus they had no greater need than did regular physicians for a knowledge of botany.[44] Medical botany did, however, remain a recommended, if less frequently required, study for regular physicians.[45] As late as 1893, no less a figure than the prominent Philadelphia physician S. Weir Mitchell counseled young doctors to take up botany or some other natural-history hobby, but his reason—to ward off boredom—reflected how nonessential botany had grown to the day-to-day activities of medical practice.[46]

Furthermore, changes in demography and in agricultural practices decreased the widespread need for knowledge of the agricultural applications of botany. Ironically, the closing decades of the nineteenth century saw renewed efforts to find and disseminate agricultural applications of science. State boards of agriculture, the new U.S. Department of Agriculture, and agricultural experiment stations championed botany by promoting horticultural applications and encouraging state agricultural schools to provide well for botany departments.[47] The work of experiment stations stressed research and dissemination, both of which encouraged the study of botany.[48] The botany that agricultural experts promoted, however, came in the form of physiology, biochemistry, genetics, and applied fields—a far cry from systematics, the bread and butter of amateurs.

More and more, the professional reaction to claims of utility was that of Brown University professor William Whitman Bailey, who expressed vexation at questions of practicality. While he enumerated a variety of applications, he stressed that science need not have any immediate, obvious applications to be legitimate, because it might be of use someday.[49] Whether or not botany was directly useful seemed not only hard to prove, but no longer worth proving. Botany was a science, and science was ultimately useful in its own right. With professionals no longer promoting botany as a useful body of knowledge for the general public, a strong leg of the utilitarian argument became shaky.

9 The Triumph of Professionalization

In the years following the Civil War, two changes occurred that would have a dramatic impact on amateur botanizers: the historic patterns of information flow that had kept amateurs within the botanical community eroded, and the type of science pursued by amateurs was no longer that pursued by the mainstream of professionals. As botany became increasingly professionalized, the interaction *between* professionals and amateurs was first supplemented and then replaced by interaction *among* professionals and *among* amateurs. This development was by no means confined to botany, but rather was part of the normal process of professionalization, occurring across the disciplines: increasingly, those who saw themselves as professionals sought to set themselves apart and to establish their social position by preempting information and by claiming expertise. Where once the distinction between communicating with the public and communicating with other experts had been gray, the new professionals made much of the distinction and valued the latter form of communication highly.[1]

By 1875, professional botanists had enough identity and sufficient numbers to create their own journals and societies. The *Botanical Gazette*, for example, not only was written and edited by professionals, but consciously advanced the ideal of professional autonomy and new, distinctly professional research interests. In the final quarter of the century, institutions that were clearly aimed at *either* amateurs *or* professionals but *not* both became increasingly common, culminating in the creation of the professional Botanical Society of America in 1892. The division was never total: anyone could subscribe to the *Botanical Gazette*, and many amateurs did; the professional domination in writing and editing and the new biological orientation, however, left little doubt as to whose journal it was. By 1900, the exchange networks that had served earlier generations so well were no longer viable. Amateurs and professionals operated in separate spheres. The impact of this separation can only be measured in contrast with the earlier years of cooperation and collaboration, starting with the first attempts to organize botany.

The leadership of professionals was not confined to high places. The third quarter of the century saw American scientists of all disciplines increasingly concerned with professionalization and with improving the quality of American science. In local societies, just as in national groups like the American Association for the Advancement of Science (AAAS), this took the form of an increased tendency for professionals to be the leaders. Authorities, and professional positions for them to fill, were no longer as rare as they had been. The Providence Franklin Society, for example, was still open to any white male, but its presidents were increasingly apt to have had at least medical training, or to be very accomplished in some area of science.[2]

New societies also tended to put professionals in positions of leadership. In the late fifties and early sixties, a small group of New Yorkers laid the groundwork for what would become the most successful botanical society in America, the Torrey Botanical Club. The club's meetings in its early years were informal gatherings with John Torrey. Eventually, a regular time was set, and in 1867 the club began to organize more formally.[3] The Torrey Botanical Club, perhaps because of its informal origins, always encouraged the participation of amateurs. Dedicated as it was to the study of the flora of the New York City area, it sponsored regular field trips from its early days. Yet it never descended to the level of a social club, because its strong leadership, always accomplished and often professional, ensured "that the club would encourage serious work and that its local interests would be pursued in more than a parochial context."[4] The place of amateurs within the club, while valued, was increasingly prescribed as the professional leadership sought to establish its authority. The Torrey Botanical Club grew rapidly in its early years, from 31 "founders" to 76 members in 1885, 114 in 1888, and 237 in 1900. It assumed its name in 1870 when one of the members took it upon himself to begin publishing a monthly four-page newsletter which evolved into the *Bulletin of the Torrey Botanical Club.*[5]

The *Bulletin of the Torrey Botanical Club* was in its early years a forum for club messages and finds. Like the club itself, the *Bulletin* gained immediate acceptance and soon became "without doubt the leading American outlet for the publication of taxonomic research."[6] It was especially useful for amateurs, because it provided a forum for the publication of minutes of local societies that could not afford their own journals. Increasingly, though, professionals found the mix within the *Bulletin* unsatisfactory and the proper standards for publication lacking. John Merle Coulter, one of the new generation of professionals, wrote in 1882 to a colleague in a rage about the publication of descriptions of new species by someone known to be careless: "Have you any influence with the *Torrey Bulletin?* What can it mean to admit new species described by *Marcus E. Jones,* in a magazine too that must be referred to?"[7]

In 1875 another, very different journal joined the *Bulletin of the Torrey Botanical Club* in the effort to spread the botanical word. The *Botanical Bulletin,* which soon became the *Botanical Gazette,* was the creation of a young Indiana professor, John Merle Coulter. Originally intended to complement the *Bulletin of the Torrey Botanical Club* by covering the Midwest, the *Gazette* rapidly became a national journal of a very different sort. The success of the two journals demonstrates the growth of the botanical community. The marked contrast in their tone and style, however, is evidence of how far professionals and amateurs had drifted apart. On the surface, both the *Bulletin* and the *Gazette* were vehicles for club reports and botanical news. The early volumes of the *Bulletin of* the *Torrey Botanical Club*, however, stressed natural history, and they were filled with notes by and about amateurs. The *Botanical Gazette*, in contrast, rapidly became a professional journal, lauding the New Botany. It was shepherded by and written for professionals.

The creator and first editor of the *Bulletin of the Torrey Botanical Club* was club member William H. Leggett, an accomplished amateur who made his living as a classicist and school principal. In its early years, the *Bulletin*'s monthly issues were filled with documentation of New York state flora and news of the Torrey Botanical Club and other clubs. The club news included not only information on new members and reports on planned meetings, but also social notes detailing the lighter side of club meetings. The *Bulletin* reported, for example, on jelly that one member had made from wild quince. The *Bulletin*'s scientific content was heavily taxonomic, reflecting a natural-history orientation. It made no overtures to professionalism or the New Botany.[8]

In the late seventies, a group of ladies from Syracuse, New York, founded a botanical club. One of their first acts was to announce the club's formation and plans in the *Bulletin of the Torrey Botanical Club*. As they grew and prospered, they continued to send regular notes to the *Bulletin*, including mention of finds and discussion of meetings.[9] Rapidly expanding to thirty-two members, the group rented rooms, sponsored field trips and classes, created a common herbarium, and held fund-raisers to pay for their expenses. Keeping the group together was not easy: what started out as a friendly rivalry among the members became near war, when one faction refused to reveal the site of an especially prized find. As early as 1878, two of the most active members, Mary O. Rust and Kate Barnes, wrote to the Harvard circle to report the club's activities and to seek help with tricky identifications. The payoff for corresponding with the Harvard group and publishing accounts of their activities in the *Bulletin* was not only help with the identification of specimens, but also letters of introduction to others in the botanical world.[10]

One of the introductions that Syracuse's botanizers wished was to the *Botanical Gazette*. Although they had found it easy to write directly to the *Bulletin of the Torrey Botanical Club*, they realized that the *Gazette* was less open to amateurs.[11] The *Gazette* rapidly became, as its editor hoped, "a necessity" to botanists, and just as

The Syracuse Botanical Club in the field, 16 July 1915
(Courtesy of the Archives of the Gray Herbarium, Harvard University)

rapidly it became evident that it was botanists, not botanizers, who were in charge. The scientific content was heavily imbued with the New Botany, an approach that had little place for amateur involvement.[12]

The New Botany

Amateur and professional interests became increasingly incompatible during the 1880s and 1890s as the dominant professional focus shifted from natural history to biology. The New Botany was a biologically oriented approach to the entire plant world. Unlike natural history, it employed experimentation as well as observation, stressing physiology and ecology over taxonomy. The New Botany did not abandon taxonomy, but deemphasized it in favor of physiological and ecological issues.[13] A new generation of professional botanists, led by Charles Edwin Bessey, John Merle Coulter, and others eager to attain professional identity and autonomy, were able to use the New Botany effectively to define a body of knowledge that they controlled to the exclusion of amateurs and other scientists.[14] Natural history became sufficiently associated with amateurism throughout the scientific community that in 1876 New York's Lyceum of Natural History changed its name to the New York Academy of Sciences—not merely to reflect its increasing breadth, but also because younger leaders feared that "Natural

History" implied amateurism and because "Lyceum" implied entertainment.[15]

The New Botany was a laboratory-oriented approach to botany that eschewed the old natural-history approach in favor of a more modern biological focus on physiology and ecology. Unlike field botany, the New Botany required specialized training and equipment, making it inaccessible to many amateurs. Professionals used the New Botany to institutionalize and to develop professional autonomy. J. M. Coulter's *Botanical Gazette* paid token respect to amateurs but was distinctly professional. The *Gazette*'s enthusiasm for the New Botany subtly pervaded the pages of every issue. That enthusiasm is most obvious in Coulter's characterization of Bessey's *Botany for High Schools and Colleges* as "not only the *best* American book on the subject, but the *only* one."[16] Professionals perceived the new approach as giving botany "greater scientific authority," which in part meant that it placed the science firmly in the grasp of professionals.[17] Only professionals had the expertise to execute the New Botany, and certainly only they could judge it. By focusing their journals and societies around the New Botany, professionals set themselves apart from amateurs.

The *Botanical Gazette*'s monthly editorial made its professional approach abundantly clear, offering advice to amateurs on what to do and how to do it. In 1880, a long editorial chastised amateurs for "spending much valuable time in getting together material that has already been collected, or is not important enough to justify the trouble"; they compounded this sin when they then insisted on sending the material on in the form of "trivial letters" to prominent professionals, who were "compelled to submit to such impositions in the hope of gleaning now and then a few grains of wheat from all this chaff." The remedy, suggested the editorial, was for amateurs to form clubs through which work could be directed by professionals, and for club leaders to send notable finds to journal editors, who could then publish what was worthy.[18] The message should have been obvious to any amateur: this journal represented the interests of professionals.

The New Botany was riddled with problems for amateurs. The most fundamental was that it required costly equipment and laboratories, and extensive training as well. Equally, if not more, important was the inability of the New Botany to meet the needs and interests of amateurs: it fit neither the old motivations—

natural theology, gentility, and utility—nor the newer ones—exercise and fun—and it provided no new ones to fill the gap. It dismissed natural theology as unscientific, and it offered little obvious in the way of self-improvement. It failed to satisfy the love many amateurs had of collecting and assumed instead an experimental urge that many amateurs lacked. Pedagogically, it was unworkable at the precollege level. In short, amateurs did not adopt the New Botany because unlike natural history it offered them nothing. It served only professional interests.

The reaction of amateurs to the New Botany and the *Gazette*'s assertions was understandably cool. The *Gazette* periodically published letters from amateurs and their defenders, arguing that while the New Botany was a fruitful approach for professionals, it lacked interest for amateurs and was a pedagogical failure. The more common response on the part of amateurs was simply to ignore the New Botany and the professionals. Faced with rejection from professionals, they sought information from other sources and networks.

As the spheres of professional and amateur botany became increasingly distinct, the *Gazette* and the *Bulletin* represented the pinnacle of botanical publishing, but other sources on botany abounded. During the years following the Civil War, American magazines flourished. Simultaneously, many existing magazines showed an interest in science for the first time and many new magazines incorporated it. The *Atlantic*, which had included occasional scientific pieces since its founding, formally incorporated science in 1868. Edward L. Youmans used his editorial discretion at *Appleton's* in the sixties and seventies to fill the journal's pages with both science and fiction with scientific themes. The *Atlantic* and *Appleton's*, like other popular journals, portrayed botany and botanizing as solely the collection and identification of plants firmly cast in the natural-history mold.[19]

New journals devoted to science also flourished. Joining the old *American Journal of Science* were the *American Naturalist* (1867), *Popular Science Monthly* (1872), *Science* (1883), and a host of lesser journals.[20] These periodicals had diverse purposes and readerships. The *American Naturalist* was originally conceived as a "popular illustrated magazine of natural history," but by the early seventies it had adopted a more sophisticated stance that included a special interest in evolution.[21] Its articles, tailor-made for amateurs, included practical suggestions on preserving specimens, advice for the urban botanist, and news of governmental and society involvement. Eventually, however, it became "so scientific and so technical that no one but a specialist can get anything out of it that is helpful to him."[22] *Science*, in contrast, was destined to become the voice of the American Association for the Advancement of Science in 1900, and it had a sophisticated professional outlook from the start. *Popular Science Monthly*, perhaps the most important of the three new magazines for amateurs, was founded in 1872 by Edward L. Youmans. Although Youmans himself had no special interest in botany, his sister Eliza did, and she contributed regularly to the *Monthly*. Edward Youmans's mission, as he saw it, was to make the works of great scientists accessible to the public. Hence, the pages of *Popular Science Monthly* revealed more interest in promoting than in advancing science.[23]

The distinction between the promotion and the advancement of science proves extremely useful for understanding the two spheres of the late-nineteenth-century botanical community. Seen clearly in periodicals, the distinction is even more obvious in societies. The new societies of the seventies and later were, like the Syracuse Botanical Club, increasingly composed of amateurs *or* professionals, as opposed to amateurs *and* professionals. Even more explicit was the emergence in the 1880s of two professionally oriented societies: The Botanical Society of America, and the Botanical Club of the AAAS, which evolved into Section G. At the head of both were the leaders of the New Botany, including Bessey and Coulter, who saw the New Botany as a tool for "raising the standard of American botanical science."[24]

In 1883, the botanical members of the AAAS banded together to form the American Botanical Club; nine years later this club spawned the Botanical Society of America, which limited its membership to "those who have published worthy work and are actively engaged in botanical investigation." By imposing both scholarly publication and research as membership requirements, the Botanical Society of America effectively confined itself to professionals. Indeed, controversy existed even among professionals during the early years about the extremely tight standards for membership that the Society imposed. When the meeting of the American Botanical Club voted to establish the Society it elected ten men who, on the basis of publication and current research,

"beyond all question should belong to a society so restricted."
Those ten met immediately and selected an additional fifteen "who
in their judgement fall well within the limits suggested." Those
twenty-five were then charged with maintaining the standards of
membership, among other things. As one of the twenty-five wrote:

> The aim of the society was stated to be the promotion of
> botanical research, and it was thought best that the limits on
> membership should be very rigidly drawn. It will, I think, be
> evident that the original list of twenty-five members contains
> the names of no persons not entitled to membership on the
> basis of creditable published work. . . . The object being the
> promotion of research it was thought best not to make mem-
> bership, at least in the charter list, complimentary to distin-
> guished botanists who are no longer engaged in the perfor-
> mance of active work, hence, the omission of certain honored
> names which must suggest themselves to every American
> botanist. For the same reason, it was the sense of the mem-
> bers present that continued membership in the society should
> depend on the continued activity of members and a continued
> interest in this principal aim of the society.[25]

The notion that amateurs might serve some useful scientific
purpose for professionals had declined with the death of Asa Gray,
the wane of natural history, and the rise of the New Botany. Those
professionals who did see a continued role for amateurs stressed
that that role was a subordinate one. Coulter's comments in the
Botanical Gazette exemplify the new professional conviction that
the exchange between professionals and amateurs was largely one-
way, consisting of professionals advising amateurs. The *Botanical
Gazette*, and even the *Bulletin of the Torrey Botanical Club*, offered
just such advice, either from more-experienced amateurs or from
professionals who sought to educate and direct amateurs. This
reshaping of the role of amateurs was not something confined to the
botanical community: other branches of what had been natural
history all faced it, as did those aspects of other sciences where
amateurs had played a lively role. In astronomy, for example,
where amateurs had been essential helpers in observation, watch-
ing the sky with far more eyes than the professional community
could muster, the issue came to a head with the formation of the
American Astronomical Society in 1899. The decision there was

to include amateurs, in what Marc Rothenberg has described as an overt, deliberate attempt by professionals to control them and channel their energies in useful directions.[26]

Late-century surges in professionalization and the advent of the New Botany resulted in breakdowns in communication between professionals and amateurs, even in informal institutions like clearinghouses. Because the New Botany deemphasized taxonomy, its adherents were less dependent upon pools of collectors. Some professionals suggested that amateurs take up new tasks, such as careful observation of the life cycles of plants—but they were explicit in insisting that such work must be done under the direction of professionals, for professional use. And even those professionals who advocated a role for amateurs complained that few amateurs provided any real scientific help. Some candidly admitted that they saw amateurs as bores and inconveniences, whose only useful function was the financial support of professional science.[27]

In light of these social and scientific threats to amateur botany and the coincident flourishing of professionals, it would be easy to accept at face value the applicability of Nathan Reingold's observation that the "great trend" of science in America from 1820 to 1920 is the disappearance of "cultivators," or avocational scientists.[28] Botanical amateurs certainly did disappear from view in professional societies and journals. Increases in the numbers of professionals, and the changes in the style of research associated with the professional embrace of the New Botany, largely eliminated the scientific need for the work of amateurs in any form but that of patrons and audiences. In gaining autonomy, professionals won the ability to redefine "science" and "scientist." As far as professionals were concerned, amateurs were no longer members of the scientific community, and in that sense they did cease to exist. The era of a botanical community composed of workers of all levels of expertise, commitment, and training was over.

It would be a mistake, however, for historians to accept the professional botanists' view that the lack of amateur participation in the New Botany led to the *disappearance* of amateurs. Indeed, an interest in nature, botanizing included, flourished in the late nineteenth century. Americans devoted new energy to preserving wilderness and wildlife from encroachment by urbanization and industrialization. Money flowed into museums of natural history and other institutions. New natural-history periodicals appeared at record rates. Popular magazines included botany and other areas of natural history with new enthusiasm. Books on nature—both nonfiction and fiction—sold in record numbers. Interest in natural history was thriving in 1900. This enthusiasm opened a new era in which botanizing was more closely allied with hobbies than with professional science. The predominant form that the hobby of botanizing took in 1900 was Nature-Study.[29]

During the eighties and nineties, professionals in the vanguard of botany, especially Charles Edwin Bessey, grew increasingly concerned about the generally poor level of botany teaching. Bessey represented the new generation of American botanists who were university-trained and laboratory-oriented.[30] Even in the wake of Gray's attempt to introduce more physiology, much school botany continued to be field-oriented. Bessey firmly believed that botanical education should begin in the laboratory, with physiological and microscopical work. Through botany he sought to teach scientific thinking, with no interest in by-products such as exercise or natural theology. He did not abandon fieldwork: he claimed to send his own students out as early in the spring as weather permitted; he did, however, deemphasize fieldwork, preferring disciplined laboratory study.[31] Bessey's own laboratory-oriented *Botany for High Schools and Colleges* (1880) was designed for use with a flora appropriate for the locale.[32] More than simply a pedagogical reform, this represented an attempt to impose a professional way of thinking and professional ideals upon amateurs. In doing so it ran head-on into a pedagogical program that was designed with amateurs in mind, the Nature-Study movement. As a scientific reform, the New Botany was a success. As a pedagogical movement, the results were, as we shall see below, less favorable.

As the cutting edge of professional botany turned away from the natural history of plants, it turned from a program that worked best when amateurs were heavily utilized, to the biology of plants, a program with little room for amateur involvement. This was not simply a case of professionals cutting amateurs out of the action: amateurs generally found the New Botany unappealing, as it minimized the possibilities for self-improvement or for fun. As leading professionals lost interest in what amateurs had to say, and vice versa, the nature of the interaction between the two groups, not surprisingly, changed. The relationship that had served both groups so well collapsed. Amateurs and professionals alike found that their needs were served best by societies, journals, and informal information exchanges that catered to either one group or the other. Talking less and less to each other, the two groups grew into separate spheres moving in different directions.

The reshuffling of botanical networks was hardest on individuals like J. G. Lemmon and Increase Allen Lapham who were intermediates between amateurs and professionals. The amateur who collected for professionals became an endangered species, at home neither in the newly defined botanical community of professionals nor in the amateur Nature-Study movement. As professionalism and biology thrived, amateurs found a new home in the Nature-Study movement, and professionals managed nicely without collecting networks. Without the support of professionals, however, botanizers found it much harder to gain proficiency and expertise or even to get necessary supplies and information. Seen through the eyes of professionals, amateur botany had failed to adjust to professionalization and the modernization of botany. The view from the botanizers' perspective tells a different story. True, botanizers and botanizing were no longer the valued segment of the scientific community that they once had been. What botanizing had lost scientifically it had, however, made up for socially. Botanizing met the twentieth century on its own terms as a thriving hobby outside the view of science (and, unfortunately, of many of its historians). Science's loss was Nature-Study's gain.

10 The Nature-Study Movement: The Legacy of Amateur Botany

While Charles Bessey and other professional botanists advocated the New Botany as both a scientific and a pedagogical reform, another coalition of educators devised a very different approach to teaching science: Nature-Study.[1] During the 1880s and 1890s some Americans, especially middle- and upper-class urbanites, idealized the rural and rustic. As they began attempting to beautify cities and preserve the country, they realized to their horror that urban children knew very little about nature. One study in 1880 of middle-class children from Boston who were about to enter school revealed that 75 percent had no concept of the seasons, 80 percent could not identify even the most common trees, and 90 percent were unaware of the origins of cotton and leather.[2] Concurrently, Cornell University's Liberty Hyde Bailey and other agricultural educators concluded that rural Americans' lack of understanding of nature was a major contributing factor in the agricultural depression of the 1890s. Bailey reasoned that if farm children learned to appreciate nature, they would grow up to become better farmers.[3] Those concerned with urban children also saw an urgent need for introducing nature into the curriculum. From these two forces arose the Nature-Study movement, a nature-awareness program aimed primarily at youngsters.

To the modern observer, Nature-Study and the New Botany appear to have been incompatible approaches to teaching botany—and had they been intended for the same audience, this would have been true. However, Nature-Study targeted primary schools (and, to a far lesser extent, adult extension), while the New Botany aimed at high school and college.[4] Proponents of Nature-Study did not claim to be teaching science. Writing of Nature-Study in 1903, Liberty Hyde Bailey stated: "Nature-study is not science. It is not knowledge. It is not facts. It is spirit. It is concerned with the child's outlook on the world."[5] Bailey's rationale for teaching both science and Nature-Study further reflected his view that the two were not competitors for the same audience, but, rather, contrasting approaches suited for different age groups: "Nature-study is a revolt from the teaching of mere science in the elementary grades. In teaching-practice, the work and the methods of the two intergrade, to be sure, and as the high school and college are approached, nature-study passes into science-teaching, or gives way to it."[6]

When, in 1892, the National Education Association's Committee of Ten reported on secondary-school studies, the Conference on Natural History stressed the appropriateness of Nature-Study and the New Botany for different groups of students. The Conference maintained that pupils from grades one through eight should have at least two thirty-minute Nature-Study sessions per week, integrated with art and language studies. High-school students, in contrast, were to receive laboratory work. Bessey and fellow New Botanist John Merle Coulter convinced other Conference members to endorse botany over zoology as the preferred high-school study. The high-school botany course was to be a year-long, five-day-a-week laboratory course.[7]

Despite the Committee of Ten's endorsement, the New Botany was not destined to become the dominant form of high-school botany. Although professional botanists applauded it, educators found it somewhat unworkable at the high-school level.[8] Few schools had the microscopes and laboratory facilities, or the trained teachers to instruct students in their use.[9] Further, the New Botany did not offer many of the benefits of natural history or Nature-Study, such as physical exercise and training in systematic thinking. Students and teachers alike found it overwhelmingly complex and boring:

> High school pupils of tender years were faced at the start with the compound microscope and preparations of minute micro-organisms and complex transverse sections unrelated to anything before heard of in the pupil's experience, and the connection, if there was any, with life interests of any sort was either overlooked or purposely omitted. The subjects frequently carried the incubus of long and unfamiliar Latin names. Lengthy and laborious laboratory exercises with the requirement of drawings which in the end showed little comprehension of the subject resulted in the removal of botany from the condition of familiar interest to a status conceived of as foreign and remote from ordinary human affairs. It might have its application but nobody seemed to know just how, or when, or where.[10]

As a result of these and other difficulties, those high schools that did include botany in the curriculum continued to favor taxonomically oriented field courses, rather than courses in the New Botany. As we saw in table 7, Asa Gray's texts, once the first choice of scientists, remained in the classroom as the first choice of teachers. Teachers, though ill equipped to teach the New Botany, felt comfortable using Gray.[11]

Although the New Botany never really caught on at the high-school level, Nature-Study did enjoy a successful reign. Teachers and students were far more comfortable observing bean sprouts, making a leaf collection, or cataloguing a school yard's weeds through the pedagogical approach of Nature-Study, than studying cells through the microscopes of the New Botany. This division is reflected in the expectations of colleges about the preparation of applicants. As late as the 1890s less than 25 percent of American colleges expected applicants to have studied any botany, and only 75 percent offered botany as a course of study.[12] Teachers still found that an occasional walk to "botanize" was both pedagogically valid and a welcome break. Carl Sandburg recalled a teacher who won his heart and that of his classmates in a Galesburg, Illinois, school in the 1880s by taking them for botanizing walks on nice days.[13] This tradition was too valuable to abandon easily. Thus Nature-Study, not the New Botany, shared the high-school curriculum in the late nineteenth and early twentieth centuries with general science courses, designed to meet the needs of large numbers of students efficiently, and increasingly with general biology courses.[14]

The differences between the goals of scientists and of educators as witnessed in the shift from the Linnaean texts to the natural system texts in the 1830s, and in the conflict between Asa Gray and Alphonso Wood in the 1840s, had grown still wider with the failure of the New Botany to enter the high-school curriculum. Bessey, once enthusiastic about writing for high schoolers and training their teachers, no longer felt able to communicate his expertise to the general public. His discouragement, and that of his peers, eroded the already weakened communication links between professional and amateur botanists. For much of the century, secondary-school botany had provided a vital channel between professionals and amateurs, but as the twentieth century dawned, pedagogical and scientific goals became increasingly difficult to reconcile. As high-school botany and the work of professionals grew less and less similar, amateurs and professionals lost what had been perhaps their strongest link.

Nature-Study was especially popular in elementary schools, becoming the presumed mode of instruction. When Margaret Emerson Bailey—the daughter of William Whitman Bailey, who had been a botanical advisor to the Nature-Study movement— wrote in a work of autobiographical fiction about an attempt by her mother to set up a kindergarten, Nature-Study figured prominently. Her parents struggled over curriculum when Whit Bailey refused to "instruct" the children, arguing instead that the children should teach themselves: "These days, children are supposed to grow the plants they learn about."[15] When Bailey introduced his own daughter to natural history, he used classic Nature-Study methods, giving her leaves to draw and study and a terrarium to observe, and focusing on the child's questions rather than delivering lectures.[16] Tips for teachers cautioned against too much instruction: "The teacher should studiously avoid definitions, and the setting of patterns," and "The best teacher is one whose pupils the farthest outrun him," were not untypical comments.[17] Children were to be encouraged to investigate and observe for themselves.

The *Nature-Study Review*, the primary organ of the movement, often included lesson plans for teachers to employ in their own classes. The October 1911 issue, for example, included several articles on studying weeds. A teacher from the William McGuffey School in Miami, Ohio, reported that seventh and eighth graders had begun their study with a discussion of the two things they knew about weeds: that weeds are undesirable, and that they are very successful. The question that interested the students most was how to control weeds. Toward this end the class developed four problems to explore: "(1) Identification, (2) Reproductive Potentiality, (3) Seed Dispersal, (4) Means of checking the rapid distribution of seeds." For the work on identification, students brought specimens and used manuals to identify them. Hand lenses and the herbarium of the local Agriculture Department came in handy at times. Thirty-two weeds were studied, and their seeds were labeled and stored for later study. Students were struck by the large number of seeds each plant produced, and they exercised their mathematical skills by estimating the number of seeds on a plant and the space the plant took up, then the number and space taken if each seed produced a plant the next year, and again if each of those produced a plant the following year. This exercise led the students to realize that many seeds do not mature, and to explore why. They also studied means of dispersal. Finally, they discussed ways of checking the spread of seed, buying samples of clover seed and sorting the contents of their samples into good seed, broken seed, dirt, and weed seed. They identified the weeds, based on their earlier study, and consulted with authorities to discover that the 90 percent purity of their samples was considered good. For the spring, the class planned to study different means of eradication.[18]

Other articles in the same issue of the *Nature-Study Review* discussed how and why to start a school weed herbarium, useful resources, and an additional lesson plan that focused on the wild morning glory. This latter lesson emerged from a school where the plant was overrunning the children's garden and the author cautioned the reader not to forget that the teacher's goals of "teach[ing] the children to recognize the wild morning-glory; to understand its economic status; to devise and apply a method of control; to teach its characteristic habits of life" would not be the same as the children's much simpler goal of "get[ting] rid of the weed which is crowding the other plants and over-running the garden."[19]

A lesson plan on evergreens shows many of the same trends. Suggestions for an opening lesson included a simple key for distinguishing the major groups and a list of questions designed to lead the children to observe: "How do evergreen trees grow?" "Do evergreen trees shed their leaves?" "What is a cone?" "How many scales are there in the cone you are studying?" This was followed by similar lessons on the major types of evergreens, and then by tips on how to integrate the study of evergreens with history, geography, arithmetic, and English. Students studying evergreens might, the author suggested, enjoy reading James Russell Lowell's "To a Pine Tree," or visiting a very old tree as part of a study of local history.[20]

The striking thing about these lesson plans, and others like them, is the degree to which they were driven by the students' interest. Which weeds or evergreen cones would students bring? What would they observe? What would they want to know about them? Nature-Study was intended to teach children to look at the world around them. Botany was a vehicle for that exploration, rather than an end in itself. The process was far more important than the content.

The Agassiz Association

Spreading from the schools, Nature-Study also affected the treatment of natural phenomena, including plants, in books and periodicals for children, and eventually in literature for adults. While the New Botany attempted to instill professional ideals in amateurs, Nature-Study was unabashedly amateur in approach. As amateurs increasingly felt distanced from professionals, they teamed with educators to foster extracurricular interest in Nature-Study. In 1875, amateur botanists were heavily involved in founding the Agassiz Association, the most ambitious amateur natural-history undertaking in American history. Harlan H. Ballard, a Massachusetts schoolteacher, spearheaded the creation of the Agassiz Association to encourage the study of nature. It grew rapidly, and by the 1890s it linked together more than 20,000 members, mostly children, in 1,200 chapters throughout the country. The chapters were usually small, often single families, and covered an array of areas of natural history. They were linked into regional assemblies and the national society through their publications.[21]

The Agassiz Association's choice of publishing organs was consistent with its amateur focus. In the November 1880 issue of *St. Nicholas*, the country's most popular children's magazine, Ballard wrote about the club and announced the creation of a *St. Nicholas* branch of the Association.[22] In the years that followed the Association would publish its proceedings successively in *St. Nicholas*, the *Observer*, the *Swiss Cross*, *Santa Claus*, and *Popular Science News*, all designed to reach prospective members. The Agassiz Association sections of these journals were filled with testimonials from educators, reports from chapters, queries from members, and offers for exchange of specimens.

The response to Ballard's original article in *St. Nicholas* was swift and strong. Several months later he reported: "The plan proposed in the November *St. Nicholas* of organizing a Natural History Society is meeting with unexpected favor. More than two hundred boys and girls have sent their names to be enrolled as members of the 'St. Nicholas Branch'; and 'chapters,' containing each from four to twenty members have been started in many cities and towns. Still every mail brings letters full of eager questioning. Our Lenox Chapter has been obliged to resolve itself into a committee of the whole for the purpose of answering these interesting letters."[23] Subsequent issues of *St. Nicholas* contained reports and letters. In the third report a contest was announced, with a prize of a book on collecting and pressing flowers to go to the boy or girl who sent in the six best specimens of pressed, mounted, and labeled wildflowers.[24] A later report printed a number of letters from experienced naturalists offering to help members on specific topics, and a request from a college botany class for correspondents willing to exchange specimens from their region for those collected by the class.[25] By September 1884 the report was able to announce the formation of the 683d chapter, and a convention of chapters to be held in Philadelphia.[26]

These conventions and assemblies of chapters proved highly popular. The Massachusetts State Assembly of the Agassiz Association met annually for a number of years. In 1888 the first meeting drew 150 members and guests for three days of talks and scientific sight-seeing in Boston. At the third meeting, in 1890, inspirational and scientific talks were complemented by reports from the seventeen chapters represented. The large Boston (B)— Barton Chapter reported twenty-one botanical outings (free to members of the chapter, twenty-five cents for members of other chapters, and fifty cents for the public). Other chapters also expressed interest and activity in botany, with concerns about space in which to store the chapter herbarium being a common topic.[27]

On several occasions the Agassiz Association ran correspondence courses in botany. In the early eighties a course ran through several months of *St. Nicholas*. Each issue contained instructions on specific things the student was to find and either collect or sketch; the first month's exercise, for example, focused on different kinds of roots. The specimens or sketches were to be sent to the organizer of the course for evaluation. Only one student actually followed through the entire course and received a certificate of satisfactory completion, but one suspects that more readers either completed only some of the months or simply didn't bother mailing the specimens.[28]

Later courses were far more sophisticated. In 1894 the *Observer*, which had replaced *St. Nicholas* as the organ of the Agassiz Association, announced four separate botany courses: Elementary Botany; Trees; Compositae; and Ferns and Fern-Allies. For a fee of $1.75 for members, or $2.00 for nonmembers, the student in the elementary botany course received a set of numbered specimens, blank forms for analysis, and printed lessons. The lessons required the student to read on a specific topic, then study and describe in writing and through drawings one or more of the prepared specimens. As each lesson was corrected and returned from the organizer of the course, a new one was sent out. The final lesson of the six focused on learning to identify common plants. In true Nature-Study form, the student was expected to learn largely from personal observation, with the printed lessons intended to raise questions and to guide rather than to provide information. Students were cautioned against reading textbooks on botany while taking the course.[29]

The botanizers of the Agassiz Association also linked themselves into two specialized chapters, the Asa Gray Memorial Botanical Chapter and the Linnaean Fern Chapter, each of which spawned independent societies that outlived the Agassiz Association. These chapters were organized through correspondence, unlike most Agassiz Association chapters. The Gray Memorial Chapter was organized in late 1887 or early 1888 by members specifically interested in botany who wished to formalize their existing correspondence network.[30] The Linnaean Fern Chapter originated in 1893.[31] During the early years, both groups circulated reports by mail, each member receiving a packet several times annually and sending it on to another member within a week, with his or her report on recent activities or specimens for viewing added in. Even though both chapters were broken up into small regional divisions, the difficulty of enforcing rapid and reliable distribution, coupled with the cost of mailing bulky packets, proved impractical. One member told of an incident when, in order to save money, he had dropped the packet through the mail slot at the address of the next recipient only to discover that that member had moved and the house was vacant. Members were supposed to resign when they found they could not keep up with the pace, but enforcement was lax, and before long each chapter adopted a newsletter format.[32]

The Linnaean Chapter soon left the Agassiz Association fold and began attracting more advanced workers. Professionals with a special interest in ferns were welcome; they joined because their specialty was too nascent to support a professional group. The Fern Chapter functioned not unlike the old clearinghouses for a subdiscipline that lagged behind in professionalization. They did manage to meet as a group in 1898 and again in 1900, with twenty members attending each meeting (at which papers were read and displays of specimens presented), but it was difficult to keep the group going.[33]

One of the last strong shows of natural theology occurred in the Agassiz Association, which took as one of its two mottos "Per Naturam Ad Deum," or "Through Nature to God." The Association's fostering of careful scientific work by amateurs is noteworthy, but its nonscientific aims and aspirations are no less interesting. Not the least of these was promoting middle-class values, which had traditional Protestantism as their backbone. The choice of "Per Naturam Ad Deum" was a symbol to all that the Agassiz Association was a respectable organization: what more obvious sign of conformity could one have found than the promotion of a love of God among young people? At the same time, the choice of

The fields and forests! Far better for the young folks than the crowded, distracting city.

But intelligent interest and respect for wild nature should be added to "fresh air," romp, play and hearty appetites.

"The Fields and Forests" (From *The Agassiz Association*, ca. 1890, 13)

the slogan also symbolized the inherently amateur nature of the founder's conception of the Association, an adherence to values that the professional community had left behind a generation before.

And yet, despite its adherence to tradition, the Agassiz Association was in one sense very modern—namely, in the superficiality of its homage to natural theology. While there was no question of the sincerity of the motto, there was also no depth to it. Tips on how to press plants or how to organize a collection were given far more attention than evidences of design or providence. Reports from the chapters talked of how many species had been identified, not of the wisdom and power of their Creator. Natural theology was no longer a tool for explanation, but merely an ideal, an underlying set of assumptions, even for amateurs.

The Agassiz Association was by no means the only Nature-Study club. Small, unaffiliated groups sprang up throughout the country, both in reality and in fiction. In Seattle, in Chautauqua, and elsewhere children banded together to explore nature. As Scouting emerged, quickly followed by Campfire and other similar groups, Nature-Study was adopted as an integral part of the program.[34] This embrace of Nature-Study was possible because it was, unlike the New Botany, infinitely accessible to amateurs. At its simplest, Nature-Study required only an observer and a subject. No one was too poor or too isolated to be involved. More importantly, Nature-Study appealed to amateurs by placing the participant at the center of activity. Experts acted as resources and clearinghouses, not as policy makers or directors. Nature-Study was the pursuit of personal satisfaction, not, as one professional had described amateur botany in 1880, a service to science.[35] Specimens were widely available, as well as being neat and easy to study.

Botany proved as appealing to Nature-Study patrons as it had been to antebellum Americans. The Nature-Study movement took a deliberately amateur, natural-history approach to the natural world. Botany was a favorite subject of Nature-Study enthusiasts for many of the same reasons that it had thrived earlier: good observational skills, an interest in nature, and respect for the world around us could all be taught from plants. This was in essence a conservation of the major theme of the amateur interests. Nature-Study also appealed to the new needs of amateur botanists. Advocates stressed, for example, that Nature-Study could and should be fun and exciting. The botanical activities of Nature-Study organizations included field trips and clubs that were able to compete with bicycle rides and baseball teams. Within Nature-Study, botanizing was a first-class hobby, not second-class science. The integration of botanizing into Nature-Study signaled the dawn of a new era, both scientifically and socially.

Conclusion

As the nineteenth century drew to a close, amateur botany was in transition. Social changes had weakened many of the traditional motivations for amateurs to botanize. Simultaneously, a host of new recreations supplanted botanizing as one of a handful of respectable pastimes. Amateurs and professionals found that their increasingly different goals left them with few common interests. The scientific goals of professionals had evolved in directions that amateurs found difficult or undesirable to pursue. Professionals used this split to fortify their own institutional autonomy. These changes in the social and scientific structures of botanizing spelled the end of one era of botanizing and the beginning of a new one.

Socially, amateur science held a very different place in 1900 than it had in 1830. Many of the motivations that had traditionally compelled amateurs to botanize were in decline. Turn-of-the-century promoters of culture still valued and endorsed self-improvement, natural theology, and the careful use of leisure time, but with considerably less ardor than their antebellum forebears. Late in the nineteenth century, Americans learned to play, and the number and variety of recreational opportunities exploded. These new diversions cut sharply into the potential ranks of amateurs in all sciences. In part, the new hobbies succeeded by appealing to the same sorts of motivations that amateur science had: athletics, for example, claimed to build strong bodies and characters. While Jacob Abbott's Rollo had chosen botany from among a rather narrow range of possible and, more importantly, respectable pastimes, Louisa May Alcott's late-century children fit botanizing in between learning circus tricks, dancing, and, most modernly, baseball. With such enticing competition, botanizing was bound to experience some decline in popularity.[1]

Not all of the decline was due to external forces, however: much of it came from inside the scientific community, as botany became a profession and a discipline. Part of what was involved in the emergence of botany as a discipline was a change in the organization of knowledge, which resulted in categories being rearranged and relabeled to fit new understandings of scientific thought. Like professionalization, this specialization involved the drawing of intellectual lines, certification, training, involvement in research, and the like. The profession and its subject matter are by no means unrelated, and just as historians have confused the two meanings of professional—expert and worker—we have also confused the processes of professionalization and discipline formation.[2] Changes in categories of knowledge and attempts to create autonomy and authority often go hand in hand. Distinguishing between the effects of professionalization and discipline formation, and discerning how the two are related, is often no easy feat, and the case of botany is no exception.

The shift from artificial to natural systems of classification and the shift from natural history to biology—the two major shifts in the organization of botanical knowledge in the nineteenth century—demonstrate the growing distance between amateurs and professionals that resulted from increasingly divergent motivations. In both cases the push to change came from professionals, and served the purpose—whether intentionally or unintentionally—of increasing professional control over botanical knowledge.

When Asa Gray and John Torrey, the leading botanists of the United States, sought to introduce the new systems of classification that were gaining popularity in Europe to their American colleagues in the 1830s, they did so because they saw the new natural systems as being better science than artificial systems. But for most amateurs the "goodness" of the science was a relatively minor concern, secondary to the facility with which one could identify a specimen. Artificial systems—especially the widely used Linnaean system—served the needs of amateurs very well. The result was a conflict over which system was better, in which amateurs and professionals proceeded to talk past each other for a generation. While it was certainly the case that some amateurs felt that the adoption of a natural system "takes botany from the multitudes, and confines it to the learned," the far more common reaction was to acknowledge that both systems had merit and that their goals were very different.[3] Few amateurs argued that scholarly works should retain the artificial system, but many argued that natural systems were inappropriate for everyday use and especially for teaching purposes. While the result of the introduction of natural systems may have been a tightening of professional authority, it is less clear that professionalization was the subject of the fight.[4]

Later in the century, when biology supplanted natural history as the dominant professional approach to studying the plant world, the shift was much subtler, although ultimately it was a more revolutionary transformation. Until the late 1870s American botanists, professional and amateur alike, were overwhelmingly dedicated to taxonomy. While the methods and details involved changed over the course of the century, the basic activities remained much the same: procuring, identifying, preserving, and exchanging specimens. What unified their activities was a natural-history focus. When, in the fourth quarter of the nineteenth century, the vast majority of professionals turned from natural history to biology, amateurs found that natural history, which at first had aligned them with the mainstream of professionals, now separated them from it.

With the advent of the new biological approach of professionals, which emphasized development and internal processes over classification and structure, the natural-history focus on taxonomic issues was replaced by an ecologically and physiologically dominated biology that gave new direction and definition to botany. Increasingly, professional botanists focused on narrow questions, developing high levels of expertise and communicating within small elite circles, rather than with the general public.[5] Whether one regards the transition as a Kuhnian paradigm change or simply as a major shift in priorities, it is clear that in 1910 botany involved a set of ideas, activities, questions, and answers very different from those of 1810. In his study of the emergence of professional social science, Thomas Haskell argued: "A social thinker's work is professional depending on the degree to which it is oriented toward, and integrated with the work of other inquirers in an ongoing community of inquiry."[6] By providing such an orientation for inquiry, the New Botany served as an organizing tool for the professionalization of botany.

Amateurs, whose interests lay not in the advancement of science but in personal achievement, were less able or willing to make the transition from natural history to the New Botany. The New Botany emphasized laboratory work, greatly lessening professionals' need for amateur collectors. The exercise incurred was nonexistent or limited, even in the case of ecology where the field became a laboratory. Natural theology was similarly not part of the new professional agenda. And as the microscope and other equipment became essential rather than merely desirable, the ease and inexpensiveness of botany declined.

The New Botany became the body of knowledge and techniques in which only professional botanists were expert, giving them the authority and autonomy that distinguished them both from amateurs and from other professionals in the life sciences. Only professional botanists were members of the "ongoing community of inquiry" delineated by the New Botany. It defined the discipline in a way that promoted professionalization and discouraged amateur participation. Thus what appears on the surface to have been a change in the organization of knowledge is also a key issue in professionalization, whereby professionals effectively displaced amateurs from the botanical community.[7]

As long as botanical interests were largely taxonomic, professionals and amateurs worked well together because there was enough overlap in their work to make communication fruitful. However, as professionals increasingly took to the laboratory, the agendas of professionals and amateurs diverged so much that cooperation became less important and less profitable. Professionals were no longer as dependent on amateurs for critical mass or scientific assistance. On its own and as part of Nature-Study, botanizing in 1900 was of little interest to professionals. Amateurs in turn found the New Botany less able to meet their needs and turned to other hobbies, such as bicycling and baseball. Natural history retained a vital hold on amateurs; biology never acquired it.

Scientists and historians of science regrettably seem to have assumed that because amateurs were no longer active participants in the scientific community, they no longer existed. Thus the case of botany illustrates the great importance of the lessons of the new social and cultural history of the past several decades. When the professionalization of botany is studied solely through the eyes of professionals, amateurs lose importance as the story progresses because they ceased to contribute to the advancement of science as professionals triumphed. While that is not wrong, it is also not the whole story. When we understand that amateurs viewed the experience very differently, and were not trying to contribute, we come to see them not as inadequate professionals but as a different sort of being altogether. We need to remember when we look at amateurs not to define them in contrast to professionals—that is, in apposition to workers and experts—but rather in their own right—as lovers of science. Only then will we remember that they sought very different ends, and only then will we come to understand what happened when science became a profession.

Notes

INTRODUCTION

1. Davis, *Life and Work of Cyrus Guernsey Pringle*; Davis, "Pringle Herbarium"; Brainerd, "Cyrus Guernsey Pringle."

2. Daley, "Vermont Fern Expert."

3. Veysey, "Who's a Professional?"

4. For an example, see Ainley, "Contribution of the Amateur." Two notable exceptions are Kohlstedt, "Nineteenth-Century Amateur Tradition"; J. H. Warner, "'Exploring the Inner Labyrinths.'"

5. See, for example, Ross, "Scientist"; Perry, *Intellectual Life*.

6. Kohlstedt, "Nineteenth-Century Amateur Tradition," 175.

7. *Oxford English Dictionary*, s.v. "amateur," "professional"; Stebbins, "Amateur"; Stebbins, *Amateurs*; Bledstein, *Culture of Professionalism*, 31–32.

8. Freidson, *Professional Powers*, 21–26; J. Brown, "Professional Language," 33–35.

9. Freidson, *Professional Powers*, 21–26.

10. Haskell, *Authority of Experts*, xii.

11. Scholars are divided on the issue of whether those who acquire this expertise are able to do so because they are favorably situated in society, or whether they use the acquisition to gain power and authority, which then places them in a favorable position in society. See Haskell, *Authority of Experts*, xx–xxiv; in the same volume, see Magali Sarfatti Larson, "The Production of Expertise and the Constitution of Expert Power," 28–80; Eliot Freidson, "Are Professions Necessary?," 3–27.

12. Wilensky, "Professionalization of Everyone?"

13. For two examples, see Ben-David, "Science as a Profession," 874; J. H. Abrams, "Professionalism in Public Organizations."

14. Cravens, "American Science Comes of Age," 53; Daniels, "Professionalization."

15. Reingold, "Definitions and Speculations," 34.

16. For one review of the literature, see Geison, introduction to *Professions*.

17. Many did, however, practice medicine or hold a college teaching position that involved some science. See Elliott, "Models," 85–86.

18. Reingold, "Definitions and Speculations."

19. C. C., "Study of Botany."

20. For a contrasting interpretation, see Daniels, "Professionalization," 162–63.

1. Atran, "Origin of the Species and Genus Concepts," 216–17.
2. Rudolph, "Introduction of the Natural System."
3. Quoted in Savage, *Discovering America*, 65.
4. Wilson, "Dancing Dogs," 225–27.
5. Ibid.
6. Waterhouse, *Botanist*, vi; T. C. Johnson, *Scientific Interest*, 47.
7. Smallwood, "Amos Eaton," 171.
8. Reingold, "Definitions and Speculations," 62.
9. McAllister, *Amos Eaton*, 229.
10. Lincoln Phelps, *Botany for Beginners* (1833); Lincoln Phelps, *Familiar Lectures* (1829); *National Union Catalogue*, 454:562–68.
11. Lyon, "Centennial," 484.
12. For a survey of the literature on literacy see Kaestle, "History of Literacy."
13. For a guide to the quantitative studies see Elliott, "Models."
14. "Death of an Old Botanist"; Deane, "Thomas Morang."
15. Flagg, "Botanizing," 658.
16. Loveland, "Half-Mile Walk."
17. W. W. Bailey, "About Weeds."
18. McAtee, "Journal of Benjamin Smith Barton," 94–95.
19. Shepard, "Botany" (1891).
20. E. I. Fernald, "Michael S. Bebb," 15.
21. Gray, *Botanical Text-Book*, 514.
22. "Collection and Preservation of Plants"; Flagg, "Botanizing," 660.
23. Thoreau, *Journal*, 2:1091.
24. W. W. Bailey, *Botanizing*, 11–12.
25. Gray, *Elements of Botany for Beginners and Schools*, 184; Gray, *Botany for Young People and Common Schools*, 99.
26. Feith, "Summer Vacation," 43–45.
27. Gray, *Botany for Young People and Common Schools*, 99. Advertisements for lenses and other equipment were common in journals that catered to botanizers. See, for example, *Observer* 1 (1890): 4; 7 (1896): 460.
28. Gray, *Botany for Young People and Common Schools*, endpage; Gray, *Elements of Botany for Beginners and Schools*, 187.
29. W. W. Bailey, *Botanizing*, 15.
30. Ibid., 17–18.
31. Asa Gray to Charles Bessey (?), 23 November 1888, Hunt Institute for Botanical Documentation, Carnegie-Mellon University.
32. W. W. Bailey, *Botanizing*, 18.
33. Roorbach, *Bibliotheca Americana*, passim.
34. W. W. Bailey, *Botanizing*, 22.
35. Ibid., 21–22.
36. M. E. Bailey, *Good-Bye, Proud World*, 30–32; W. W. Bailey, *Botanizing*, 10–13.

37. "Austin, Coe Finch, 1831–1880," typescript biographical sketch in the archives of the Academy of Natural Sciences of Philadelphia, 5.

38. W. W. Bailey, *Botanizing*, 11.

39. Ibid., 10.

40. Ibid., 12.

41. Treat, "Botany for Invalids," 39.

42. Kate Furbish to Asa Gray, 25 June 1882, Historic Letter Files, Gray Herbarium, Harvard University.

43. See, for example, "Plants for Sale"; "Florida Plants."

44. Gray, *Elements of Botany for Beginners and Schools*, 184.

45. Roessler, "Outdoors."

46. See, for example, "Collection and Preservation of Plants," 351–52; Barry, "Pressing Flowers."

47. See, for example, Curtis, "Hints on Herborizing," 258.

48. See, for example, ibid., 257; Gray, *Elements of Botany for Beginners and Schools*, 184–85.

49. Gray, *Botanical Text-Book*, 514–15.

50. See, for example, Curtis, "Hints on Herborizing," 257.

51. Gray, *Botanical Text-Book*, 515.

52. Gray, *Elements of Botany for Beginners and Schools*, 184–86. Also see Curtis, "Hints on Herborizing."

53. W. W. Bailey, "Botanists' Winter Evenings."

54. E. I. Fernald, "Michael S. Bebb," 14–15.

55. M. D. R., "Summer Day's Pastime"; Jacob Abbott, *Rollo's Museum* (1839), 137–53; C. C. Abbott, "Harry's Museum."

CHAPTER TWO

1. Exact counts are often impossible because membership records for botanical societies are at best fragmentary. However, even the nineteenth century's most esteemed botanical society, the Torrey Botanical Club, was overwhelmingly amateur. Barnhart, "Historical Sketch"; Burgess, "Work of the Torrey Botanical Club"; Sloan, "Science in New York City," 44–46.

2. The literature on the scientific institutions of nineteenth-century America includes Bates, *Scientific Societies*; Oleson and Brown, *Pursuit of Knowledge*; Kohlstedt, *Formation*.

3. Dupree, *Asa Gray*; Haygood, "Spheres of Influence."

4. Among the few studies that consider amateurs are Hendrickson, *Arkites*; Kohlstedt, "Nineteenth-Century Amateur Tradition."

5. Bates, *Scientific Societies*; Hendrickson, *Arkites*; Kohlstedt, *Formation*; Oleson and Brown, *Pursuit of Knowledge*.

6. For two recent reflections on the 1840s as a turning point see Bruce, *Launching of Modern American Science*; Kohlstedt, "Parlors, Primers, and Public Schooling."

7. Stearns, *Science in the British Colonies*, 708–11; Hindle, *Pursuit of Science*, 11–35.

8. Hindle, *Pursuit of Science*, 11–35; Berkeley and Berkeley, *Life and Travels of John Bartram*, 126–47 and passim.

9. Hindle, *Pursuit of Science*, esp. 127–45; M. L. Fernald, "Some Early Botanists."

10. Bates, *Scientific Societies*, 51; Meisel, *Bibliography of American Natural History*, vols. 1–2 passim.

11. "Notice of the Late Dr. Waterhouse"; Gerstner, "Academy of Natural Sciences of Philadelphia," 174, 177.

12. Greene et al., *Providence Plantations*, 211–12; W. O. Brown, *Providence Franklin Society*.

13. Ramsay, *History of South Carolina*, 2:60–61. Curiously, in 1817–18 an unrelated course of popular botany lectures in Charleston proved highly successful: Shecut, *Medical and Philosophical Essays*, 44–47.

14. Mott, *American Magazines*, 1:215, 341–42.

15. On the history of periodicals in nineteenth-century America see Mott, *American Magazines*. On scientific journalism see Beaver, "Altruism"; Whalen and Tobin, "Periodicals."

16. Mott, *American Magazines*, 1:302–4; Baatz, " 'Squinting at Silliman.' "

17. Hawkes, "Increase A. Lapham," 78.

18. *American Journal of Science* 42 (1842).

19. Goodale, "Development of Botany." Also see Daniels, *American Science*, 38–39.

20. "Art. II.—Decandolle's Botany," *North American Review* 32 (1834): 32–63.

21. For an example of a botanizer who felt threatened see Daniels, *American Science*, 38–41.

22. A characteristic example of this view is Lincoln Phelps, *Familiar Lectures* (1851), 4 and passim.

23. This trend was not confined to botany. For a wide-sweeping example see Kohlstedt, *Formation*.

24. Ibid., 200–204.

25. Quoted in Kohlstedt, "Nineteenth-Century Amateur Tradition," 177 (italics in original).

26. Among the best accounts is Dupree, *Science in the Federal Government*, 66–90.

27. S. F. Baird to Ferdinand Lindheimer, 8 March 1851, Outgoing Correspondence of the Assistant Secretary, 1:234, Smithsonian Institution Archives; S. F. Baird to Charles Wright, 4 January 1851, ibid., 1:144–46. Also see Deiss, "Spencer F. Baird"; Dupree, *Science in the Federal Government*, 66–90.

28. Hindle, *Pursuit of Science*, 224–25.

29. A. D. Rodgers, *John Torrey*.

30. See, for example, John Russell to Increase Allen Lapham, 17 December 1857; T. Hale to I. A. L., 20 September 1859, both in the Lapham Papers.

31. Hawkes, "Increase A. Lapham," 78.

32. S. F. Baird to John Torrey, 5 October 1859, Outgoing Correspondence of the Assistant Secretary, 20:34, Smithsonian Institution.

33. S. F. Baird to Ferdinand Lindheimer, 8 March 1851, Outgoing Correspondence of the Assistant Secretary, 1:234, Smithsonian Institution; S. F. Baird to Charles Wright, 4 January 1851, ibid., 1:144.

34. The correspondence between Lapham and Gray indicates a lively exchange of specimens, especially in the years when Gray was working on distribution. See, for example, Asa Gray to Increase Allen Lapham, 27 October 1958, Lapham Papers. For a discussion of this exchange see Hawkes, "Increase A. Lapham," 136.

35. Dupree, *Asa Gray*, 391–92 and passim.

36. J. G. Lemmon to Asa Gray, 3 August 1875, Historic Letter Files, Gray Herbarium.

37. J. G. Lemmon to Asa Gray, 21 December 1881, ibid.

38. J. G. Lemmon to Asa Gray, 30 October 1875, ibid.

39. J. G. Lemmon to Asa Gray, 5 March 1878, ibid.

40. Rebecca M. Austin to George Davenport, 22 September 1877, ibid.

41. J. G. Lemmon to Asa Gray, 6 December 1877, ibid.

42. Gray's articles, notes, and reviews in the *American Journal of Science* are too numerous to list, appearing in nearly every issue. His most noted pieces in the *Atlantic Monthly* were a series on Darwinism, which are reprinted in Gray, *Darwiniana*.

43. *Godey's Lady's Book* 58 (1859): 562.

44. *Atlantic Monthly* 4 (1859): 29–40; 6 (1860): 257–70.

45. See, for example, "Agricultural Botany."

46. See, for example, "Notices of New Publications," *Youth's Cabinet* 1 (1846): 291; Theodore Thinker, "Familiar Conversations on Botany," which were regular features starting in vol. 2 (1847) and continuing on into the 1850s.

47. Elliott, "Models," 86.

48. Sloan, "Science in New York City"; Kohlstedt, "Nineteenth-Century Amateur Tradition." The essays in Oleson and Voss, *Organization of Knowledge*, and Oleson and Brown, *Pursuit of Knowledge*, present a number of case studies. An analysis of the historic roots of the current role of amateurs is given in Meadows and Fisher, "Gentlemen V. Players." Another science and another country are examined in Porter, "Gentlemen and Geology."

CHAPTER THREE

1. Alphonso Wood, *Class-Book of Botany* (1860), 12.

2. Veysey, *Perfectionists*; Thomas, "Romantic Reform"; Walters, *American Reformers*.

3. Curti, *Growth of American Thought*, 346–48, 360–73.

4. "Moral Uses of Plants," 41.

5. Noll, *Botanical Class-Book*, v.

6. Lorimer, *Among the Trees*, 28, recalling the antebellum era.

7. "Moral Uses of Plants," 43.

8. Ibid.

9. "Botany for Schools," 172.

10. "Lessons in Botany—No. II," 91.

11. Epis. Obs., "Young Naturalists."

12. Recounted in Browne, "Botany," 66.

13. Green and Congdon, *Analytical Class-Book*, iv. See also Epis. Obs., "Young Naturalists"; Flagg, *Studies*, 1–2.

14. Clark, *Relations of Botany to Agriculture*, 4.

15. "Economy and Habits of Plants," 24.

16. Coultas, *What May Be Learned from a Tree*, 12.

17. On one manifestation of many of the components of gentility, see Welter, "Cult of True Womanhood."

18. Noll, *Botanical Class-Book*, v. Similarly, see "Botany"; Riddell, "Brief Sketch," 449; "Lessons in Botany," 66; Tilgate, "Botany."

19. Lincoln Phelps, *Fireside Friend* (1855), 272.

20. Flagg, *Studies*, 1–7.

21. See, for example, C. H. Turner, *Floral Kingdom*; or M. W. D., "Spring Flowers."

22. C. C., "Study of Botany," 106. See too "Botany for Schools," 172.

23. Coultas, "Wild Flowers of Early Spring Time," 343. See also "Botany for Schools," 171–72; Allen, *Naturalist in Britain*, 229–31; Kastner, *World of Watchers*; Gibbons and Strom, *Neighbors to the Birds*.

24. Tilgate, "Botany."

25. C. C., "Study of Botany," 106.

26. "Lessons in Botany—No. II," 91. Similarly, see Browne, "Botany," 66; "Study of Botany in Common Schools"; *Girls' Manual*, 178.

27. Katz, *Early School Reform*, 135.

28. "Botany as a Study for Young Ladies." See too "Flowers"; Thinker, *First Lessons in Botany*.

29. Emerson, "Education of Females," 29; Gould, "Introduction of Natural History," 227–28.

30. "Lessons in Botany—No. II," 91. See also Lincoln Phelps, *Familiar Lectures* (1831), 10, 14; Lincoln Phelps, *Botany For Beginners* (1837), 10; "Botany for Schools," 170–71; D[avid] Hosack quoted in Eaton, *Manual of Botany for the Northern and Middle States* (1824), vii–ix; Bolzau, *Almira Hart Lincoln Phelps*, 219–22.

31. Browne, "Botany," 66. On the importance of orderliness see D. T. Rodgers, *Work Ethic*.

32. Lincoln Phelps, *Fireside Friend* (1855), 85.

33. "Lessons in Botany—No. II," 91. See also Wakefield, *Introduction to Botany*, 17.

34. Coultas, "Hints to Ladies Studying Botany."

35. *Girls' Manual*, 191. See also Alphonso Wood, *Class-Book of Botany* (1860); Lincoln Phelps, *Fireside Friend* (1855), 56; "Lessons in Botany—No. II," 91.

36. "Botany for Schools," 169–70. See too C. C., "Study of Botany," 106–7; "Letter of P. R. Hoy"; Noll, *Botanical Class-Book*, v.

37. "Lessons in Botany—No. II," 91.

38. Veysey, *Perfectionists*; Thomas, "Romantic Reform."

39. McDonald, "Autobiographical Notes," 22.

40. Harrington, *Analysis of Plants*, 1.

41. Sherwin, "Relative Importance," cited in Katz, *Early School Reform*, 136.

42. "Sketch of Prof. Gray," 490. See also Higgins, "Vestiges"; "Editors Table"; King, *Scheme*.

43. "Lessons in Botany—No. II," 91. See also Youmans, *Descriptive Botany*, x; Emsby, "Systematic Botany Nevertheless," 298–300. In a similar vein see J. F. A. Adams, "Is Botany a Suitable Study for Young Men?"; "Sketch of Prof. Gray," 490; King, *Scheme*.

44. Youmans, *First Book of Botany*.

45. Youmans, *Descriptive Botany*, x.

46. J. F. A. Adams, "Is Botany a Suitable Study for Young Men?" See also Whorton, *Crusaders for Fitness*, esp. 92–131, 270–330.

47. "Testimony of George B. Emerson," 133–34.

48. Amos Eaton wrote to John Torrey in 1816 that his wife Sarah had been ill: "She does no business of consequence only to collect and preserve plants, and teach the children" (quoted in McAllister, *Amos Eaton*, 168). Nor was this phenomenon confined to America: see Allen, *Naturalist in Britain*, 131–32.

49. E. F. Andrews, "Botany as a Recreation for Invalids," 779.

50. Ibid., 780–81. See too Brotherton, "Notes on Michigan Cypripediums," 382–84.

51. E. F. Andrews, "Botany as a Recreation for Invalids," 781.

52. Kate Furbish to G. G. Davenport, 14 May 1893, Historic Letter Files, Gray Herbarium. See too Treat, "Botany for Invalids."

53. E. F. Andrews, "Botany as a Recreation for Invalids," 781.

54. E. S. Phelps, *Doctor Zay*, 102–3.

55. Alcott, *Eight Cousins*, 148; Alcott, *Under the Lilacs*, 125–31.

56. See, for example, Brainerd, "Cyrus Guernsey Pringle"; "Obituary: Isaac Burk, Botanist," clipping dated 31 March 1893, in Francis Whittier Pennell, "Biographies of Botanists," Academy of Natural Sciences of Philadelphia; M. E. Jones, "How I Became a Botanist," MS in the Hunt Institute for Botanical Documentation, Carnegie-Mellon University.

57. Alcott, *Under the Lilacs*, 125–31.

58. M. D. R., "Summer Day's Pastime."

59. "How to Converse"; Harris, *Wild Flowers*, 68–69.

60. Other interesting examples include Dodge, *Louise and I*, 198–215; "Walking-Fern." Additionally, M. E. Bailey, *Good-Bye, Proud World*, presents a fascinating autobiographical view of the child of a society woman and an academic botanist at the turn of the century.

61. Spaulding, "Botany in the High School."

CHAPTER FOUR

1. Nineteenth-century Americans used the term "seminary" to denote a school for females between the ages of about twelve and twenty-one, on an academic level somewhat equivalent to a modern high school. The term "academy" usually referred to a similar school for males, though it occasionally referred to a female or coeducational endeavor.

2. Curti, *Growth of American Thought*, 335–57; Church and Sedlak, *Education*, 55–83.

3. Guralnick, *Science*, 22, 109–11.

4. Ibid., 22.

5. G. F. Miller, *Academy System*, 108–17.

6. Inglis, *Rise of the High School*, 73.

7. For two of the many discussions see C. Bode, *American Lyceum*, 41–98; Bates, *Scientific Societies*, 28–84.

8. Based on an examination of seminary and academy catalogues at the Maryland Diocesan Archives, held in trust at the Maryland Historical Society, and the Library of Congress.

9. Hayner, *Troy and Rensselaer County*.

10. On Rensselaer see Reznick, *Education*, 4–78, 484, 496. On Troy Female Seminary, now the Emma Willard School, see Lutz, *Emma Willard*, esp. 83–116; Fairbanks, *Emma Willard*.

11. "Lesson in Botany," 254.

12. On Pestalozzi see Silber, *Pestalozzi*; Barlow, *Pestalozzi and American Education*. See too McAllister, *Amos Eaton*, 384–93, 485–86; Scott, "Almira Hart Lincoln Phelps," 208; Smallwood, "Amos Eaton," 181–83; Aldrich, "New York Natural History Survey," 23–27; "Lesson in Botany," 254.

13. For a discussion of these sorts of arguments see Lincoln Phelps, *Botany for Beginners* (1837), 9–12.

14. See Guralnick, *Science*, 28–33; Guralnick, "Sources of Misconception."

15. Sherwin, "Relative Importance," cited in Katz, *Early School Reform*, 136; "Botany for Schools," 170–71.

16. See Bolzau, *Almira Hart Lincoln Phelps*, 68–85, 209; "Mrs. Almira Lincoln Phelps," 613–14; Rudolph, "Almira Hart Lincoln Phelps"; McAllister, *Amos Eaton*, 205 and passim.

17. *National Union Catalog*, 154:364.

18. Ibid., 454:562–63, 568.

19. Ibid., 564–67.

20. "Review of Lincoln's *Familiar Lectures on Botany*"; *Godey's Lady's Book* 23 (1841): 23–24.

21. For example, see "Untitled Note," Marcus E. Jones Papers, Box 4, Folder "Manuscripts, Autobiographical" in Hunt Institute for Botanical Documentation; Geary and Hutchinson, "Mr. Dawson, Plantsman," 52.

22. For a nonbotanist's perspective see "Review of Lincoln's *Familiar Lectures on Botany*."

23. See, for example, Lincoln Phelps, *Fireside Friend* (1840), 190–91; Lincoln Phelps, *Familiar Lectures* (1851), 4.

24. Rudolph, "Introduction of the Natural System," 461–68; Aldrich, "New York Natural History Survey," 104–8.

25. Lincoln Phelps's most explicit endorsement was in her "Popular Botany," 288–89. For the views of like-minded others, see Thinker, *First Lessons in Botany*, iii; "Botany for Schools," 171.

26. Lincoln Phelps, *Familiar Lectures* (1851), appendix, 545.

27. McAllister, *Amos Eaton*, 237–38 (italics Eaton's).

28. Letter to John Torrey, 27 December 1820, quoted in McAllister, *Amos Eaton*, 228.

29. Lincoln Phelps, *Familiar Lectures* (1851), 4; Eaton, *Manual of Botany for the Northern and Middle States* (1833), iv–vii.

30. Even Troy Female Seminary charged an extra fee ($3.00 per term) after Lincoln Phelps's departure. See *Catalogue of the Officers and Pupils*, 3.

31. Among the many is Angell, *Angell's Fifth Reader*, 36–38 and passim.

32. Laura, "Letter from a Correspondent." For a somewhat later example see Sandburg, *Always the Young Strangers*, 114, writing of his childhood in the 1880s.

33. Gray's experience is described in Dupree, *Asa Gray*, 19–55. For information on other physician-botanists see Kelley, *Some American Medical Botanists*.

34. In contrast, American colleges did not, as a rule, add botany until 1840–45. See Guralnick, *Science*, 109–11.

35. Dupree, *Asa Gray*, esp. 52–54.

36. Ibid., 48, 52–54, and passim.

37. *American Gardener's Magazine* 2 (1836): 421–24; *American Journal of Science* 30 (1836): 399. For references to individuals converted to the natural system by Gray, see Dupree, *Asa Gray*, 53, 63. Also see Rudolph, "Introduction of the Natural System."

38. Gray, *Botanical Text-Book for Colleges*; Eaton, *Manual of Botany for the Northern and Middle States* (1833); Lincoln Phelps, *Familiar Lectures* (1851).

39. Gray, *Elements of Botany*, vi.

40. O. R. Willis, "Biographical Sketch of Dr. Alphonso Wood, of West Farms, Professor of Botany in the College of Pharmacy, New York," read before the Torrey Botanical Club, n.d., n.p., copy in the Smithsonian Institution Botany Library Biographical Files; Bacon, "Prof. Alphonso Wood."

41. "Classbook of Botany," 190.

42. Dupree, *Asa Gray*, 170–71.

43. Asa Gray to John Torrey, 15 November 1845, in Gray, *Letters*, 1:334–36.

44. Dupree, *Asa Gray*, 170.

45. Asa Gray to John Torrey, 21 November 1845, in Gray, *Letters*, 1:336.

46. Dupree, *Asa Gray*, 170–73.

47. Gray's texts included: *First Lessons in Botany and Vegetable Physiology* (1857); *Botany for Young People and Common Schools. Part I: How Plants Grow* (1858); *Botany for Young People. Part II: How Plants Behave* (1872). Wood's included: *First Lessons in Botany* (1843); *Leaves and Flowers: or, Object Lessons in*

Botany (1860); *The American Botanist and Florist* (1870); *Fourteen Weeks in Botany* (1879).

48. G. F. Miller, *Academy System*, 108–9.

49. Stout, *Development of High School Curriculum*, 153. Stout studied the "North Central" States: Illinois, Indiana, Iowa, Kansas, Michigan, Minnesota, Missouri, Nebraska, North Dakota, Ohio, South Dakota, and Wisconsin.

50. Gray, *First Lessons*, 178, 195; Gray, *Botany for Young People and Common Schools*, 96. While Lincoln Phelps was a vocal advocate of natural theology, she never addressed it from this angle.

51. "Annual Report of the Regents of the University," State of New York Senate Document 4, 1890, 633.

52. Tobey, *Saving the Prairies*, 29.

53. *Botanical Gazette* 6 (1881): 168.

54. Tobey, *Saving the Prairies*, 24–47.

CHAPTER FIVE

1. Eaton, *Manual of Botany for the Northern and Middle States* (1824), ix.

2. McAllister, *Amos Eaton*, 442.

3. C., "Flowers."

4. "Study of Botany in Common Schools"; T. H. Johnson, *Complete Poems of Emily Dickinson*, untitled poem no. 70, 36–37.

5. J. F. A. Adams, "Is Botany a Suitable Study for Young Men?" See also D. J. Warner, "Science Education for Women"; Rudolph, "How It Developed"; Rudolph, "Women."

6. Shteir, "Linnaeus's Daughters"; Shteir, "Botany in the Breakfast Room"; Rudolph, "How It Developed"; Rudolph, "Women."

7. Quoted in H. W. Rickett, "Jane Colden as Botanist in Contemporary Opinion," in Rickett, *Botanic Manuscript*, n.p.

8. Wilson, "Dancing Dogs."

9. Ibid.; and see, for example, *Notable American Women*, 1:357–58.

10. D. J. Warner, "Science Education for Women."

11. Welter, "Cult of True Womanhood."

12. Flagg, "Botanizing," 657–58.

13. C., "Flowers." See also Lincoln Phelps, *Familiar Lectures* (1831), 14.

14. C. W., "Genera Florae," 444. See also Wakefield, *Introduction to Botany*, iv–vii.

15. For several different views of how educators and moralists interpreted and used the notion of true womanhood, see Melder, "Mask of Oppression"; Scott, *Making the Invisible Woman Visible*, esp. 37–106; Riley, *Inventing the American Woman*, 63–119; Hersh, "'True Woman'"; Smith-Rosenberg, *Disorderly Conduct*.

16. Scott, *Making the Invisible Woman Visible*, 92.

17. Coultas, "Wild Flowers of Early Spring Time," 343.

18. Browne, "Botany," 74. See also "Lessons in Botany—No. II," 91; Emsby, "Systematic Botany Nevertheless."

19. Mavor, *Catechism of Botany*, 3.

20. "Study of Botany in Common Schools." See also C. W., "Genera Florae," 446; Katz, *Early School Reform*, 135.

21. For example, see "Maria's Visit," 46–47; Dodge, *Louise and I*; Alcott, *Eight Cousins*, esp. 148; Alcott, *Under the Lilacs*, esp. 125–31.

22. Lincoln Phelps, *Familiar Lectures* (1831), 14. See also Lincoln Phelps, *Fireside Friend* (1840), 56; "Letter from the School Girls"; A. D. Wood, " 'Fashionable Diseases.' "

23. Betts, "American Medical Thought," 142 and passim; [Higginson], "Health of Our Girls," 728–29; Dana, *How to Know the Wild Flowers*, vii–viii; Lincoln Phelps, *Fireside Friend* (1840), 56; Wakefield, *Introduction to Botany*, iv, 1–2; Estelle, "Botany, No. I"; Coultas, "Wild Flowers," 372–73.

24. Lincoln Phelps, *Familiar Lectures* (1831), 14. In a similar vein, see Flagg, "Botanizing."

25. Betts, "American Medical Thought"; [Higginson], "Health of Our Girls"; Sklar, *Catharine Beecher*, 204–16; Verbrugge, *Able-Bodied Womanhood*.

26. Lincoln Phelps, *Fireside Friend* (1840), 55–56; Cole, *Hundred Years*, 62; E. A. Green, *Mary Lyon*, 220, 287–88.

27. Flagg, "Botanizing," 657.

28. Charlotte A. Ford, "Eliza Frances Andrews."

29. Flagg, "Botanizing," 657–58.

30. "Wild Flowers," 198–99.

31. Theodore Thinker, "Familiar Conversations on Botany," *Youth's Cabinet* 2 (1847): 135; "Botany for Schools," 170.

32. Quoted in Bolzau, *Almira Hart Lincoln Phelps*, 77.

33. Ibid., 147.

34. Eaton and Wright, *North American Botany* (1840), vi.

35. Lorimer, *Among the Trees*, 30.

36. Roessler, "Summering."

37. Cole, *Hundred Years*, 62; E. A. Green, *Mary Lyon*, 220, 287–88; Roessler, "Outdoors." Mount Holyoke was not the only school where botanizing threatened the local flora. Antioch's president in the early 1860s, Thomas Hill, cautioned students: "Do not take all the flowers from any one plant, leave most for seed. When you see a plant which needs help, give it. Do not pick plants or flowers for the mere pleasure of picking them" (quoted in Filler, *Ohio Schoolmistress*, 189).

38. T. C. Johnson, *Scientific Interest*, 112.

39. Filler, *Ohio Schoolmistress*, 189.

40. Denner, "Science and Long Skirts."

41. Wakefield, *Introduction to Botany*, 1–2.

42. M. D. R., "Summer Day's Pastime"; Dodge, "How to Prepare Sea-Mosses"; "How to Converse."

43. Alcott, *Eight Cousins*, 148.

44. Church, *Flower-Talks at Elmridge*.

45. Feith, "Summer Vacation," 26–29, 43–44. See too "Four 'Inland' Girls by the Sea."

46. "How to Converse."

47. William M. Beauchamp, *Past and Present*, 37–38; *Botanical Gazette* 4 (1879): 206–7; 5 (1880): 45. See too the letters from Mary Olivia Rust and Kate S. Barnes to Asa Gray, George Davenport, and others in the 1870s and 1880s, Historic Letter Files, Gray Herbarium, Harvard University.

48. "Editorial," *Botanical Gazette* 7 (1882): 64.

49. J. Reverchon, "Botanizing in Texas," 216. Similarly, see "Botanist in Arizona"; A. Hunter Dupree and Marian L. Gade, "Mary Katherine Layne Curran Brandegee," in *Notable American Women*, 1:228–30.

50. N. B. Turner, *Testament of Happiness*, 78.

51. For a full description of this trend see Rossiter, *Women Scientists*.

52. Setchell, "Townshend Stith Brandegee."

53. Eastwood has been the subject of many short sketches. Among the best are L. Abrams, "Alice Eastwood"; "Biographical Sketch of Alice Eastwood"; and Joseph Ewan, "Eastwood, Alice," in *Notable American Women*, 4:216–17, which includes a bibliography of other sketches.

54. Kate Furbish to Asa Gray, 25 June 1882, Historic Letter Files, Gray Herbarium, Harvard University.

55. Lazella Schwarten, "Kate Furbish," in *Notable American Women*, 1:687. See also Coburn, *Kate Furbish*; S. A. Adams, "Rare Flower"; Kate Furbish to Asa Gray, 25 June 1882, Historic Letter Files, Gray Herbarium.

56. S. A. Adams, "Rare Flower." See too Furbish's own accounts in the *American Naturalist* 15 (1881): 469–70; 16 (1882): 397–99.

CHAPTER SIX

1. D. T. Rodgers, *Work Ethic*, 14.

2. Ibid., 125–52; Hardy, *How Boston Played*, esp. 41–62.

3. D. T. Rodgers, *Work Ethic*, 94–96, 130–31.

4. Alphonso Wood, *Leaves and Flowers*, 5. See also O. R. Willis, "Biographical Sketch of Dr. Alphonso Wood, of West Farms, Professor of Botany in the College of Pharmacy, New York," reprint from unidentified source in Biographical Files, Botany Library, Smithsonian Institution.

5. C. W., "Genera Florae," 448.

6. Mitchell, "Lectures on the Conduct of the Medical Life," 655. My thanks to Dr. Bruce Fye for this citation. Also see W. W. Bailey, "Botanists' Winter Evenings"; E. F. B., "Botanizing at the Coal Bin"; "Trees in Winter."

7. Jacob Abbott, *Rollo's Museum* (1861), esp. 136–53; Lyman Abbott, "Snap-Shots."

8. Epis. Obs., "Young Naturalists." For further examples see Theodore Thinker, "Familiar Conversations on Botany," *Youth's Cabinet* 2 (1847): 136, 300; "Wild Flowers," 198–99; Coxe, *Floral Emblems*, 7–8.

9. "Maria's Visit."

10. "Lessons in Botany"; Mavor, *Catechism of Botany*, 4; Clark and Rennie, *Alphabet of Botany*.

11. Lincoln Phelps, *Botany for Beginners* (1837), 9. See also Lincoln Phelps, *Familiar Lectures* (1852) and other editions; Lincoln Phelps, *Lectures to Young Ladies*, 208.

12. *Pastime of Learning*, 3.

13. Lincoln Phelps, *Familiar Lectures* (1852), 295.

14. [Goodrich], "About the Leaves of Trees."

15. Gray, "On the Importance of the Natural Sciences," 95.

16. See, for example, Brand, *William Rollinson Whittingham*.

17. "Letter from the School Girls." Letters appeared especially regularly in the *Youth's Cabinet* and *St. Nicholas*.

18. D. T. Rodgers, *Work Ethic*, 17–22 and passim.

19. Billings, "Study of the Minute Fungi," 329. For a typical attack on the teaching of botany without specimens see "Botany for Beginners."

20. Flagg, "Botanizing," 658. For a similar sentiment see "Botanist in Arizona."

21. Alphonso Wood, *Leaves and Flowers*, 5. For a similar account see Theodore Thinker, "Familiar Conversations on Botany," *Youth's Cabinet* 2 (1847): 135–36.

22. Flagg, "Botanizing," 658.

23. B., "Amusements."

24. Delta, "Sea-Weed Album."

25. D. T. Rodgers, *Work Ethic*. On the rise of sports see Spears and Swanson, *History of Sport*; Whorton, *Crusaders for Fitness*; Harvey Green, *Fit for America*.

26. Alcott, *Under the Lilacs*, 125–31; Alcott, *Eight Cousins*, 148. See too D. T. Rodgers, *Work Ethic*, 125–52.

27. "Editorial," *Botanical Gazette* 18 (1891): 180–82.

28. *Plant World* 1 (1897–98): 29–30.

29. Hall, *Botany*, 4.

30. Ibid.; Knight, *Primer of Botany*, iii–iv.

31. *Plant World* 1 (1897–98): 29–30.

32. Alphonso Wood, *Leaves and Flowers*, 5.

33. Shepard, "Botany's Charms"; *Plant World* 1 (1897–98): 29–30; Bebb, "Some Mistaken Estimates"; Bioletti, "Experience in Herbarium Making."

34. Bebb, "Some Mistaken Estimates." Emerson quoted by Dana, *How to Know the Wild Flowers*, viii.

35. Dana, *How to Know the Wild Flowers*, vi.

36. Plimpton, *Oakes Ames*, 54–55. Eighty years later it had a similar effect on me.

37. Kate Furbish to Asa Gray, 25 June 1882, Historic Letter Files, Gray Herbarium.

CHAPTER SEVEN

1. Egan, "Bouquet of Roses," 103. See also M. E. B., "Notes on the Fungi of Maryland"; Chickering, "Flowers of Early Spring."

2. Elson, *Guardians of Tradition*, 9.

3. Ibid., 18. See also Sanders, *New School Reader*, 203.

4. Sanders, *Sanders' Union Reader*, 188.

5. Elson, *Guardians of Tradition*, 29.

6. Eaton, *Manual of Botany for the Northern and Central States* (1824), viii, quoting David Hosack.

7. Lincoln Phelps, *Familiar Lectures* (1836), 15.

8. Gray, *Botany for Young People and Common Schools*, 2.

9. *National Union Catalogue*, 438:415–35; Glick, "Bishop Paley."

10. Glick, "Bishop Paley."

11. Clarke, *Paley*; LeMahieu, *Mind of William Paley*.

12. Clarke, *Paley*; LeMahieu, *Mind of William Paley*. See too Numbers, *Creation by Natural Law*, 77–82.

13. Hovenkamp, *Science and Religion*, esp. 3–95; Bozeman, *Protestants*; Gillespie, "Preparing for Darwin"; Daniels, *American Science*, 50–55.

14. Daniels, *American Science*, 65; Hovenkamp, *Science and Religion in America*; Bozeman, *Protestants*.

15. Thomas, "Romantic Reform"; Curti, *Growth of American Thought*, 360–73.

16. I. C., "Divinity of Nature."

17. See, for example, John Muir's comment ca. 1866 in W. F. Bode, *John Muir*, 1:147. On Geology see Greene, "Science and Religion"; Guralnick, "Geology and Religion."

18. Lincoln Phelps, *Botany for Beginners* (1836), frontispiece, 5–6. See also Newman, *Boudoir Botany*, 19.

19. Gray, *Botany for Young People*, 1.

20. Ibid., 1–2.

21. "Lessons Whispered on a Walk."

22. Lincoln Phelps, *Familiar Lectures* (1836), 15.

23. Annaline, "Evidence of Design in Nature"; Mrs. H. Smith, "Season of Flowers"; "Mystery of Nature"; "Lichens."

24. McGuffey, *New Fourth Eclectic Reader*, 69–75; also see Elson, *Guardians of Tradition*, 18–19. On the history of this Washington anecdote see Weems, *Life of Washington*, xxxi–xxxii, 13–16.

25. "Botany for Schools," 169–72.

26. Gray, "On the Importance of the Natural Sciences," 117.

27. "To the Invisible *Author*." See also Newman, *Illustrated Botany*, iv; L., "Threads from the Web," 315; Thinker, "Transplanted Tree."

28. "Study of Flowers," 158.

29. "Curious Trees." Also see Browne, "Botany"; Thinker, *First Lessons in Botany*, 41–42, 120–21.

30. "Curious Trees," 14–15.

31. Darby, *Manual of Botany*, iii.

32. Theodore Thinker, "Familiar Conversations on Botany, No. III," *Youth's Cabinet* 2 (1847): 215; Lincoln Phelps, *Familiar Lectures* (1831), 14; "Rock, Tree, and Man"; Bickley, *Principles of Scientific Botany*, xxxi.

33. C., "Flowers."

34. Egan, "Bouquet of Roses"; "Moral Uses of Plants"; Alden, "Cousin Mary and the Flowers."

35. Bickley, *Principles of Scientific Botany*, 31.

36. For the role of natural law in astronomical natural theology see Numbers, *Creation by Natural Law*, 77–87.

37. Alden, "Cousin Mary and the Flowers," 28.

38. Phoebus, *Plant Life*, n.p.

39. Thinker, "Early Spring Flowers."

40. A Southerner, "Life in the Southwest," 252.

41. Phoebus, *Plant Life*, n.p. See also "Rock, Tree, and Man"; Darby, *Manual of Botany*, iii.

42. Hovenkamp, *Science and Religion*, 37–56.

43. Ibid.; Bozeman, *Protestants*.

44. Carter, *Spiritual Crisis*, 14; Hovenkamp, *Science and Religion*, esp. 37–49. For a discussion of the relation between professionalization and attitudes about natural theology in the conchology community, see Gillespie, "Preparing for Darwin."

45. Dupree, *Asa Gray*, 296–99; Gray, *Darwiniana*.

46. "Our Native Orchids."

47. Baker, "Some Resurrection Plants."

48. Baily, *Trees, Plants, and Flowers*, v–vii. See also Hovenkamp, *Science and Religion*, esp. 45–49.

49. Egan, "Bouquet of Roses," 103; M. E. B., "Notes on the Fungi of Maryland."

50. Chickering, "Flowers of Early Spring."

CHAPTER EIGHT

1. Browne, "Botany," 66; Shepard, "Botany" (1891); R. E., "Talk about Trees," 42. See too Darby, *Botany of the Southern States*, 3; Ayers, "Claims of Natural History," 118; *Pastime of Learning*, 4; Alphonso Wood, *Class-Book of Botany* (1860), 9; Lincoln Phelps, *Fireside Friend* (1855), 191; Willis, *Practical Flora*, 10; C. W., "Genera Florae," 446–47; Mavor, *Catechism of Botany*, 9.

2. C. W., "Genera Florae," 446–47.

3. Lincoln Phelps, *Fireside Friend* (1855), 73.

4. Darby, *Botany of the Southern States*, 3. See too "Study of Botany in Common Schools"; Hazzard, "Study of Plants," 7.

5. Clark, *Relations of Botany to Agriculture*, 10.

6. Allston, "Address," 118.

7. Theodore Thinker, "Familiar Conversations on Botany," *Youth's Cabinet* 3 (1848): 39–43.

8. Hawks, *American Forest*; Clark, *Relations of Botany to Agriculture*, 20.

9. See, for example, Gillman, "Our Northern Orchids."

10. Darlington, *Address*, 11–12.

11. Ibid., 10–12. See also "Agricultural Botany"; Allston, "Address," 119–20; Hazzard, "Study of Plants"; Mavor, *Catechism of Botany*, 4.

12. Newbury, "Weeds and Words."

13. See, for example, G. B. H., "Little Botanical Knowledge"; Buckley, "Best Botany."

14. Theodore Thinker, "Familiar Conversations on Botany," *Youth's Cabinet* 4 (1849): 285–86.

15. "Edible Fungi." See also Palmer, "Toadstool-Eating"; Russell, "Mushrooms"; Clark, *Relations of Botany to Agriculture*, 15–16.

16. Willey, "Lichens under the Microscope"; "Lichens," 182–83.

17. "Agricultural Botany." See also G. B. H., "Little Botanical Knowledge"; Buckley, "Best Botany"; Hazzard, "Study of Plants"; Allston, "Address," 119–20.

18. Darlington, *Address*, 10–11.

19. Davidson, "Liberal Education for Farmers."

20. True, *History*, 246–49. On Bessey see Overfield, "Charles E. Bessey"; Walsh, "Charles E. Bessey."

21. Thomson, *New Guide to Health*, 211.

22. Elias Smith, *American Physician*, 58–59.

23. Simon Abbott, *Southern Botanic*.

24. Beach, *American Practice of Medicine*, 3:20–22.

25. Berman, "Impact."

26. Lincoln Phelps, *Fireside Friend* (1855), 73. See too "Gray's *Manual of Botany* and *Flora of North America*," 179; "Art. II. Gray's Botanical Text-Books," 341.

27. "Botany for Schools," 171.

28. Theodore Thinker, "Familiar Conversations on Botany," *Youth's Cabinet* 3 (1848): 241.

29. Lincoln Phelps, *Fireside Friend* (1855), 73; C. W., "Genera Florae," 448.

30. Kaufman, "American Medical Education," 12.

31. Cowen, "Materia Medica and Pharmacology."

32. "Botany for Schools," 171.

33. "On Botany"; "Botany for Schools," 171.

34. "On Botany"; Riddell, "Brief Sketch."

35. "On Botany," 777.

36. Ibid.; Mitchell, "Lecture on the Conduct of the Medical Life," 655; Haller, *American Medicine*, 204–9.

37. Coultas, *What May Be Learned from a Tree*, 53–54.

38. "On Plants."

39. C. W., "Genera Florae," 448.

40. Coultas, "Wild Flowers."

41. Browne, "Botany."

42. For a botanical example see "Editorial," *Botanical Gazette* 12 (1887): 68. See also Daniels, *American Science*, 21–26; Reingold, "American Indifference."

43. Daniels, "Process of Professionalization."

44. Berman, "Impact," 213–316.

45. Cowen, "Materia Medica and Pharmacology"; J. F. A. Adams, "Is Botany a Suitable Study for Young Men?"

46. Mitchell, "Lecture on the Conduct of Medical Life," 655.

47. Clarke, *Relations of Botany to Agriculture.*

48. See Halsted, "What the Station Botanists Are Doing."

49. W. W. Bailey, *Botanizing,* 1–2.

CHAPTER NINE

1. On the broader phenomenon in the life sciences see Farber, "Transformation of Natural History." On the process of professionalization see Daniels, "Professionalization"; Cravens, "American Science Comes of Age"; Kohlstedt, "Nineteenth-Century Amateur Tradition."

2. Pre-1850 presidents included a merchant, a mayor, and two high-school teacher/principals; presidents in 1850–79 included one merchant who was among Providence's most experienced botanists, 4 M.D.'s, and an elusive character who was either an M.D. or a watchmaker. See Greene et al., *Providence Plantations,* 212–13.

3. Barnhart, "Historical Sketch," 12–13; Burgess, "Work of the Torrey Botanical Club," 533; Sloan, "Science in New York City."

4. Sloan, "Science in New York City," 45.

5. Ibid.; Barnhart, "Historical Sketch," 14; Burgess, "Work of the Torrey Botanical Club," 555.

6. H. A. Gleason quoted in Sloan, "Science in New York City," 46.

7. John M. Coulter to Sereno Watson, 29 March 1882, quoted in A. D. Rodgers, *American Botany,* 202.

8. Sloan, "Science in New York City," 45–46; *Bulletin of the Torrey Botanical Club* 6 (1875–79): 48.

9. *Bulletin of the Torrey Botanical Club* 6 (1875–79): 289, 330, 345–47; 8 (1881): 24.

10. Letters from Kate Barnes and Mary Olivia Rust are in the Historic Letter Files, Gray Herbarium.

11. Mary Olivia Rust to Asa Gray, 19 August 1879, Historic Letter Files, Gray Herbarium.

12. A. D. Rodgers, *John Merle Coulter,* 33.

13. Cittadino, "Ecology."

14. Ibid., 174–81; Tobey, *Saving the Prairies,* 24–47.

15. Sloan, "Science in New York City." For a different interpretation of the name change, see Baatz, *Knowledge,* 118.

16. Coulter, "Botany for High Schools and Colleges," 98.

17. Cittadino, "Ecology," 177.

18. *Botanical Gazette* 5 (1880): 36–38.

19. See, for example, "Walking-Fern"; Mott, *American Magazines,* 3:104–5.

20. On the spurt of new journal starts see Tobey, *Saving the Prairies,* 25–26.

21. Mott, *American Magazines,* 3:108.

22. Parry, "Herbarium Cases," 472. See also W. W. Bailey, "About Weeds"; "Herbarium Paper," *American Naturalist* 7 (1873): 691; C. E. Bessey, "To the Editor."

23. Leverette, "Science and Values," 49–50.

24. For a clear statement of this intention see "Editorial," *Botanical Gazette* 14 (1889): 268–69.

25. Tippo, "Early History," 3.

26. Rothenberg, "Organization and Control."

27. Farlow, "Task of American Botanists," 312–13; "Editorial," *Botanical Gazette* 5 (1880): 36–38.

28. Reingold, "Definitions and Speculations," 47.

29. H. S. Miller, *Dollars for Research*; Tobey, *Saving the Prairies*, 25; Schmitt, *Back to Nature*.

30. Tobey, *Saving the Prairies*, 35–47; Cittadino, "Ecology," 174–81; Overfield, "Charles E. Bessey," 162–69; Walsh, "Charles E. Bessey."

31. Tobey, *Saving the Prairies*, 27–32.

32. See too Coulter, "Botany for High Schools and Colleges."

CHAPTER TEN

1. I choose to capitalize and hyphenate "Nature-Study" because that was the choice of its founders, who sought to distinguish their specific pedagogical program from other study of nature. For a discussion of the merits of this linguistic convention see Minton, "History of the Nature-Study Movement," iv.

2. Schmitt, *Back to Nature*.

3. A. D. Rodgers, *Liberty Hyde Bailey*, esp. 215–19; L. H. Bailey, *Nature Study Idea*, 62–85.

4. L. H. Bailey, *Nature Study Idea*, 3 and passim; Schmitt, *Back to Nature*, 77–105.

5. L. H. Bailey, *Nature Study Idea*, 5.

6. Ibid.

7. The Committee's report is discussed and reprinted in Sizer, *Secondary Schools*, 237–38, 245. Also see Tobey, *Saving the Prairies*, 30–31; Underwood, "Study of Botany."

8. Kirkwood, "Opportunity and Obligation"; Clute, "Botany Laboratory Equipment."

9. Kirkwood, "Opportunity and Obligation," 581–82.

10. Ibid., 581. See also Underwood, "Study of Botany."

11. Tobey, *Saving the Prairies*, 29–32; *Botanical Gazette* 5 (1880): 96–97; Stout, *Development of High School Curriculum*, 153; G. F. Miller, *Academy System*, 108–9; Emsby, "Systematic Botany Nevertheless"; *Botanical Gazette* 12 (1887): 87–88, 253–54.

12. Campbell, *Biological Teaching*.

13. Sandburg, *Always the Young Strangers*, 114.

14. Tobey, *Saving the Prairies*, 31–32; Krug, *Shaping of the American High*

School, 368–70; A. D. Rodgers, *Liberty Hyde Bailey*, 214–41; Schmitt, *Back to Nature*, 77–105; Underhill, *Origins*, 155–214; Mayer, "Biology Education."

15. M. E. Bailey, *Good-Bye Proud World*, 76–77.

16. Ibid., 14–16 and passim.

17. L. H. Bailey, "Leaflet I: What Is Nature Study," 12, 14.

18. Ragland, "Weed Study."

19. *Nature-Study Review* 7 (1911): 171–77; quotation on 172.

20. A. B. Comstock, "Evergreens."

21. For a brief history of the Association's early days see "Agassiz Association."

22. Ballard, "Agassiz Association," 28–30.

23. "Agassiz Association."

24. "Agassiz Association—Third Report."

25. "Agassiz Association—Twenty-Seventh Report."

26. "Agassiz Association—Forty-First Report."

27. Concern about space in general was rampant. See *Journal of Proceedings: Massachusetts State Assembly of the Agassiz Association*, Third Annual Convention, Fitchburg, 30 May 1890, published by the Assembly.

28. "Agassiz Association—Thirty-Sixth Report." See too *St. Nicholas* 10 (1882–83): 797, 877, 957; 11 (1883–84): 78, 260–61.

29. Wight, "Our New Courses in Botany."

30. Bartlett, "History."

31. Gilbert, "Historical Sketch."

32. Bartlett, "History." For a sample list of Gray members and their activities, see "Gray Memorial Botanical Chapter."

33. Gilbert, "Historical Sketch." See too Shepard, "Botany" (1893).

34. For a good introduction see Schmitt, *Back to Nature*, 106–24. See too Dennis, *Study of Leaves*.

35. Farlow, "Task of American Botanists," 312–13.

CONCLUSION

1. D. T. Rodgers, *Work Ethic*; Nye, *Unembarrassed Muse*; Hardy, *How Boston Played*, esp. 41–62; Jacob Abbott, *Rollo's Museum* (1839); Alcott, *Under the Lilacs*.

2. Higham, "Matrix of Specialization"; Goldstein, "Foucault among the Sociologists."

3. The remark is from Chester Dewey, as quoted in Daniels, "Professionalization," 154. Daniels's interpretation is quite different from mine, however.

4. Rudolph, "Introduction of the Natural System."

5. Farber, "Transformation of Natural History"; Cittadino, "Ecology."

6. Haskell, *Emergence of Professional Social Science*, 18–19.

7. For a discussion of the introduction of biology, see Pauly, "Appearance of Academic Biology." On the importance of the rise of the university in this process, see Shils, "Order of Learning."

Bibliography

ARCHIVAL SOURCES

Baltimore, Maryland
  Maryland Historical Society
    Maryland Diocesan Archives
Cambridge, Massachusetts
  Gray Herbarium, Harvard University
    Historic Letter Files
Madison, Wisconsin
  State Historical Society of Wisconsin
    Increase A. Lapham Papers
Philadelphia, Pennsylvania
  Academy of Natural Sciences of Philadelphia
    Francis Whittier Pennell Papers
Pittsburgh, Pennsylvania
  Hunt Institute for Botanical Documentation, Carnegie-Mellon University
    Biographical Files
    Marcus E. Jones Papers
Washington, D.C.
  Smithsonian Institution Archives
    Incoming Correspondence of the Assistant Secretary
    Incoming Correspondence of the Secretary
    Outgoing Correspondence of the Assistant Secretary
    Outgoing Correspondence of the Secretary
  Smithsonian Institution Botany Library
    Biographical Files

PERIODICALS

The following periodicals were valuable sources of short articles, book reviews, editorials, letters to the editor, and other brief pieces.

American Institute of Instruction: *Lectures*, 1830–1900
American Journal of Science and Arts, 1818–1900
American Magazine of Natural Science, 1892–94
The American Naturalist, 1867–1900
The American Presbyterian Review, 1853–71
American Repertory of Arts, Sciences, and Manufacturers, 1840–42
Appleton's Journal, 1869–81
The Asa Gray Bulletin, 1896–1900

The Atlantic Monthly, 1857–1900
The Baptist Quarterly, 1867–77
Biblical Repertory, 1825–68
Bibliotheca Sacra, 1844–63
Biological Society of Washington: *Proceedings*, 1880–92
Boston Society of Natural History:
   Boston Journal of Natural History, 1834–63
   Memoirs, 1863–1900
   Proceedings, 1841–1900
The Botanical Gazette, 1875–1910
Brownson's Quarterly Review, 1844–63
Buffalo Society of Natural Sciences: *Bulletin*, 1873–1900
The Catholic World, 1855–73
Chester County Cabinet of Natural Science: *Report*, 1828–49
The Christian Disciple and Theological Review, 1813–16, 1819–21
The Christian Examiner and General Review, 1824–69
Cincinnati Society of Natural History:
   Journal, 1878–90
   Proceedings, 1876
Cleveland Academy of Natural Sciences: *Proceedings*, 1845–59
The Collector, 1882
Collector's Home Companion, 1882
The Country Gentleman, 1853–1900
Davenport [Iowa] Academy of Natural Sciences: *Proceedings*, 1867–1900
Elliott Society of Natural History: *Proceedings*, 1853–90
Erythea, 1893–1900
Evangelical Review, 1849–71
The Fern Bulletin, 1893–1912
Field and Forest, 1875–78
Godey's Lady's Book, 1830–98
Harper's Magazine, 1850–1910
Hawkeye Observer, 1885
Illinois Laboratory of Natural History: *Bulletin*, 1876–1900
Linnaean Bulletin, 1884
Lowell Offering, 1840–45
Magazine of Horticulture, Botany and All Usefull Discoveries and Improvements
   in Rural Affairs, 1835–68
Merry's Museum and Parley's Magazine, 1842–72
Natural History Club of Philadelphia: *Annual Addresses*, 1868–74
Naturalist and Collector, 1895
Naturalists' Advertisor, 1873–77
Naturalist's Companion, 1885
Naturalists' Leisure Hour and Monthly Bulletin, 1879–89
Naturalists' Quarterly, 1880
Newport [R.I.] Natural History Society: *Proceedings*, 1883–99
North American Naturalist, 1896

North American Review, 1815–1910

The Observer (published by the Agassiz Association in Portland, Conn.), 1890–97

Old Colony Naturalist, 1901–03

Philadelphia Academy of Natural Sciences: *Journal*, 1817–32

The Philadelphia Medical and Physical Journal, 1804–09

The Philadelphia Museum; or, Register of Natural History and the Arts, 1824

The Plant World, 1897–1900

Popular Science Monthly, 1872–1910

Portland [Me.] Society of Natural History:
 Journal, 1864
 Proceedings, 1862–1911

Rhodora, 1899–1910

St. Nicholas, 1873–1900

Science, 1883–1910

Scientific American, 1845–1900

Southern Agriculturist, Horticulturalist, and Register of Rural Affairs, 1828–46

The Southern Literary Messenger, 1834–64

Southern Quarterly Review, 1842–57

The Theological and Literary Journal, 1848–60

Tidings from Nature, 1884–86

Torrey Botanical Club: *Bulletin*, 1870–1910

Trenton [N.J.] Natural History Society: *Journal*, 1886–91

The Unitarian Review, 1874–78

The United States Catholic Magazine, 1843–48

United States Review and Literary Gazette, 1824–27

Valley Naturalist (St. Louis), 1880

Vermont Botanical Club: *Bulletin*, 1906–14

West American Scientist (San Diego Society of Natural History), 1884–1902

Wisconsin Naturalist: A Monthly Magazine of Natural History, 1880–91

Work and Play, 1874–76

World of Nature, 1888

Youth's Cabinet, 1837–45

The Youth's Companion, 1827–1910

PUBLISHED SOURCES

Abbott, C. C. "Harry's Museum." *Riverside Museum for Young People* 2 (1868): 348–52.

Abbott, Jacob. *Rollo's Museum*. Boston: Weeks, Jordan, and Co., 1839.

———. *Rollo's Museum*. Rev. ed. New York: Sheldon and Co., 1861.

Abbott, Lyman. "Snap-Shots of My Contemporaries: My Father—The Friend of Children." *Outlook* 129 (1929): 55–58.

Abbott, Simon. *The Southern Botanic*. Charleston, S.C.: privately printed, 1844.

Abrams, Jay Harrison. "Professionalism in Public Organizations: The Case of the Research Scientist." Ph.D. diss., State University of New York–Albany, 1975.

Abrams, Leroy. "Alice Eastwood—Western Botanist." *Pacific Discovery* 2 (1949): 14–17.

Adams, J. F. A. "Is Botany a Suitable Study for Young Men?" *Science* 9 (1887): 116–17.

Adams, Sally Aldrich. "The Rare Flower That Challenged a Dam." *Christian Science Monitor*, April 14, 1977.

Agassiz Association. *The Agassiz Association.* Stamford, Conn.: Edward F. Bigelow, 189?.

"The Agassiz Association." *St. Nicholas* 8 (1880–81): 332–33.

"Agassiz Association—Forty-First Report." *St. Nicholas* 11 (1884): 901–2.

"Agassiz Association—Third Report." *St. Nicholas* 8 (1881): 654.

"Agassiz Association—Thirty-Sixth Report." *St. Nicholas* 11 (1883–84): 502–3.

"Agassiz Association—Twenty-Seventh Report." *St. Nicholas* 10 (1882–83): 637–38.

"Agricultural Botany." *Country Gentleman* 13 (1859): 209.

Ahlstrom, Sydney E. *A Religious History of the American People.* New Haven, Conn.: Yale University Press, 1972.

Ainley, Marianne Gosztonyi. "The Contribution of the Amateur to North American Ornithology: A Historical Perspective." *Living Bird* 18 (1979–80): 161–77.

Alcott, Louisa M. *Eight Cousins.* Boston: Roberts Brothers, 1875.

————. *Under the Lilacs.* Boston: Roberts Brothers, 1878.

Alden, Professor. "Cousin Mary and the Flowers." *Youth's Cabinet* 4 (1849): 26–28.

Aldrich, Michele Alexis LaClergue. "New York Natural History Survey, 1836–1845." Ph.D. diss., University of Texas at Austin, 1974.

Allen, D[avid] E[lliston]. "Natural History and Social History." *Journal of the Society for the Bibliography of Natural History* 7 (1976): 509–16.

————. *The Naturalist in Britain: A Social History.* London: Allen Lane, 1976.

————. *The Victorian Fern Craze: A History of Pteridomania.* London: Hutchinson and Co., 1969.

Allston, William J. "An Address Delivered before the Anti-Tariff Agricultural Society of Broad River, Fairfield District [S.C.] on Its First Anniversary, in July 1829." *Southern Agriculturalist* 3 (1830): 117–20.

Andrews, Miss E. F. "Botany as a Recreation for Invalids." *Popular Science Monthly* 28 (1885–86): 779–81.

Andrews, Jane. *The Stories Mother Nature Told Her Children.* New York: C. T. Dillingham, 1889.

Angell, Oliver. *Angell's Fifth Reader: Containing Lessons in Reading and Spelling.* Philadelphia: E. H. Butler and Co., 1857.

Annaline. "Evidence of Design in Nature." *Lowell Offering* 5 (1845): 135–37.

"Art. II. Gray's Botanical Text-Books." *North American Review* 87 (1858): 321–42.

Atran, Scott. "Origin of the Species and Genus Concepts: An Anthropological Perspective." *Journal of the History of Biology* 20 (1987): 195–279.

Ayers, William O. "The Claims of Natural History as a Branch of Common School Education." *Lectures of the American Institute of Instruction* 20 (1850): 118–47.

B. "Amusements." *Country Gentleman* 1 (1853): 313.

Baatz, Simon. *Knowledge, Culture, and Science in the Metropolis: The New York Academy of Sciences, 1817–1970.* Annals, vol. 584. New York: New York Academy of Sciences, 1990.

———. "'Squinting at Silliman': Scientific Periodicals in the Early American Republic." *Isis* 82 (1991): 223–44.

Bacon, Alice E. "Prof. Alphonso Wood." *Bulletin of the Vermont Botanical Club* 3 (1908): 21–26.

Bailey, L[iberty] H[yde]. "Leaflet I: What Is Nature Study." *Cornell Nature-Study Leaflets.* Nature-Study Bulletin No. 1. Albany: State of New York, Department of Agriculture, 1904.

———. *The Nature Study Idea.* New York: Doubleday, Page and Co., 1903.

Bailey, Margaret Emerson. *Good-Bye, Proud World.* New York: Charles Scribner's Sons, 1945.

Bailey, W[illiam] W[hitman]. "About Weeds." *American Naturalist* 12 (1877): 740–42.

———. *The Botanical Collector's Handbook.* Naturalists' Handy Series, no. 3. Salem, Mass.: George A. Bates, 1881.

———. "A Botanists' [*sic*] Winter Evenings." *Field and Forest* 3 (1877–78): 57–59.

———. *Botanizing: A Guide to Field Collecting and Herbarium Work.* Providence, R.I.: Preston and Rounds, 1899.

Baily, William L. *Trees, Plants, and Flowers: Where and How They Grow.* Philadelphia: J. B. Lippincott and Co., 1870.

Baker, E. A. "Some Resurrection Plants." *Observer* 2, no. 2 (February 1891): 6.

Ballard, Harlan. "The Agassiz Association." *St. Nicholas* 8 (1880–81): 28–31.

Barber, Lynn. *The Heyday of Natural History, 1820–1870.* Garden City, N.Y.: Doubleday and Co., 1980.

Barlow, Thomas A. *Pestalozzi and American Education.* Boulder, Colo.: Este Es Press, 1977.

Barnhart, John Hendley. "Historical Sketch of the Torrey Botanical Club." *Memoirs of the Torrey Botanical Club* 17 (1917): 12–21.

Barry, Miss E. E. "Pressing Flowers." *Observer* 1 (1890): 3.

Bartlett, Harley H. "History of the Gray Memorial Botanical Association and the Asa Gray Bulletin." *Asa Gray Bulletin*, n.s., 1 (1952–53): 3–29.

Bates, Ralph S. *Scientific Societies in the United States.* 3d ed. Cambridge, Mass.: Massachusetts Institute of Technology Press, 1965.

Beach, Wooster. *American Practice of Medicine*. 3 vols. New York: privately printed, 1833.

Beauchamp, William M. *Past and Present of Syracuse and Onondaga County*. New York: S. J. Clark Publishing Co., 1908.

Beaver, Donald De B. "Altruism, Patriotism and Science: Scientific Journals in the Early Republic." *American Studies* 12 (1971): 5–19.

———. *The American Scientific Community, 1800–1860: A Statistical-Historical Study*. New York: Arno Press, 1980.

Bebb, M. S. "On Some Mistaken Estimates Made by Amateurs." *Botanical Gazette* 13 (1888): 63–64.

Beck, L[ewis] C. *Botany of the Northern and Middle States*. Albany, N.Y.: Webster and Skinner, 1833.

Ben-David, Joseph. "Science as a Profession and Scientific Professionalism." In *Explorations in General Theory in Social Science: Essays in Honor of Talcott Parsons*, edited by Jan J. Loubser et al., 874–88. New York: Free Press, 1976.

Berger, Carl. *Science, God, and Nature in Victorian Canada*. The 1982 Joanne Goodman Lectures. Toronto: University of Toronto Press, 1983.

Berkeley, Edmund, and Dorothy Smith Berkeley. *The Life and Travels of John Bartram: From Lake Ontario to the River St. John*. Tallahassee: University Presses of Florida, 1982.

Berman, Alex. "The Impact of the Nineteenth-Century Botanico-Medical Movement on American Pharmacy and Medicine." Ph.D. diss., University of Wisconsin, 1954.

Bessey, Charles E. *Botany for High Schools and Colleges*. New York: Henry Holt, 1880.

———. "To the Editor." *Guide to Nature* 4 (1911): 69–70.

Bessey, Ernst A. "The Teaching of Botany Sixty-Five Years Ago." *Iowa State College Journal of Science* 9 (1935): 227–33.

Betts, John Rickards. "American Medical Thought on Exercise as the Road to Health, 1820–1860." *Bulletin of the History of Medicine* 45 (1971): 138–52.

Bickley, George W. L. *Principles of Scientific Botany*. Cincinnati: H. W. Derby, 1853.

Bigelow, Jacob. *Florula Bostoniensis: A Collection of Plants of Boston and Its Environs*. Boston: Cummings and Hilliard, 1814.

Billings, J. S. "The Study of the Minute Fungi." *American Naturalist* 5 (1871): 323–29.

"Biographical Sketch of Alice Eastwood." *Proceedings of the California Academy of Sciences*, 4th ser., 25 (1943–49): ix–xiv.

Bioletti, F. T. "An Experience in Herbarium Making." *Erythea* 2 (1894): 31–34.

Blake, J[ohn] L. *Conversations on Vegetable Physiology: Comprehending the Elements of Botany with Their Application to Agriculture*. Boston: Crocker and Brewster, 1830.

———. *A Family Text-Book for the Country, or, The Farmer at Home*. New York: C. M. Saxon, Barker and Co., 1860.

Bledstein, Burton J. *The Culture of Professionalism: The Middle Class and the Development of Higher Education in America*. New York: W. W. Norton, 1976.

Bode, Carl. *The American Lyceum: Town Meeting of the Mind*. New York: Oxford University Press, 1956.

Bode, William Frederic. *The Life and Letters of John Muir*. Boston: Houghton Mifflin Co., 1924.

Bolzau, Emma Lydia. *Almira Hart Lincoln Phelps: Her Life and Work*. Philadelphia: Science Press Printing Co., 1936.

Boone, Weldon Wesley. *A History of Botany in West Virginia*. Parsons, W.V.: McClain Print Co., 1965.

Botanical Catechism: Containing Introductory Lessons for Students in Botany. Northampton, Mass.: T. W. Shepard, 1819.

"The Botanist in Arizona." *Botanical Gazette* 7 (1882): 8–9.

"Botany." *Observer* 1, no. 5 (May, 1890): 3.

"Botany as a Study for Young Ladies." *Godey's Lady's Book* 58 (1859): 562. Reprinted from the *North American Review* 87 (1858): 342.

"Botany for Beginners. By Maxwell T. Masters, M.D., F.R.S., Late Lecturer on Botany at St. George's Hospital. London: Bradbury and Evans." *Popular Science Monthly* 1 (1872): 370–71.

"Botany for Schools." *American Journal of Education* 4 (1829): 168–75.

Bozeman, Theodore Dwight. *Protestants in an Age of Science: The Baconian Ideal and Antebellum American Religious Thought*. Chapel Hill: University of North Carolina Press, 1977.

Brainerd, Ezra. "Cyrus Guernsey Pringle." *Rhodora* 13 (1911): 224–32.

Brand, William Francis. *Life of William Rollinson Whittingham, Fourth Bishop of Maryland*. New York: E. and B. Young and Co., 1883.

Breck, Joseph. *The Young Florist: or, Conversations on the Culture of Flowers, and on Natural History*. Boston: Russell, Adiorne and Co., 1833.

Brotherton, Wilfred A. "Notes on Michigan Cypripediums." *Observer* 4, no. 12 (December 1893): 382–84.

Brown, JoAnne. "Professional Language: Words That Succeed." *Radical History Review* 34 (1986): 33–51.

Brown, Welcome O. *The Providence Franklin Society: An Historical Address*. Providence, R.I.: The Society, 1880.

Browne, D. J. "Botany." *Naturalist* 1 (1831): 65–74.

Bruce, Robert. *The Launching of Modern American Science, 1846–1876*. New York: Alfred A. Knopf, 1987.

Bucholtz, John T. "Elisha Newton Plank." *American Midland Naturalist* 21 (1939): 523.

Buckley, S. B. "The Best Botany of the Southern States." *Country Gentleman* 16 (1860): 275.

Buckman, Thomas R., ed. *Bibliography and Natural History*. Lawrence: University of Kansas Press, 1966.

Burgess, Edward S. "The Work of the Torrey Botanical Club." *Bulletin of the Torrey Botanical Club* 27 (1900): 552–58.

Burnham, John C. *How Superstition Won and Science Lost: Popularizing Science and Health in the United States.* New Brunswick, N.J.: Rutgers University Press, 1987.

C. "Flowers." *Godey's Lady's Book* 15 (1837): 112.

C. C. "The Study of Botany." *United States Literary Gazette* 2 (1825): 103–8.

C. W. "Genera Florae Americae Boriali Orientalis Illustrata." *Southern Quarterly Review* 15 (1849): 444–48.

Campbell, John P. *Biological Teaching in the Colleges of the United States.* Bureau of Education Circular of Information, no. 9. Washington, D.C.: Government Printing Office, 1891.

Carter, Paul A. *The Spiritual Crisis of the Gilded Age.* De Kalb: Northern Illinois University Press, 1971.

Catalogue of the Officers and Pupils of the Troy Female Seminary for the Academic Year Commencing Sept. 15, 1841 and Ending August 3, 1842. Troy, N.Y.: Troy Female Seminary, 1842.

Chickering, J. W., Jr. "The Flowers of Early Spring." *American Naturalist* 3 (1869–70): 131.

Childs, Arney Robinson, ed. *The Private Journal of Henry William Ravenel, 1859–1887.* Columbia: University of South Carolina Press, 1947.

Church, Ella Rodman. *Flower-Talks at Elmridge.* Philadelphia: Presbyterian Board of Publication, 1885.

Church, Robert L., and Michael W. Sedlak. *Education in the United States: An Interpretive History.* New York: Free Press, 1976.

Cittadino, Eugene. "Ecology and the Professionalization of Botany in America, 1890–1905." *Studies in History of Biology* 4 (1980): 171–98.

Clark, Arabella, and James Rennie. *The Alphabet of Botany for the Use of Beginners.* New York: Peter Hill, 1833.

Clark, William S. *The Relations of Botany to Agriculture: A Lecture Delivered before the Massachusetts Board of Agriculture at Barre, December 9th, 1872.* Boston: Wright and Potter, 1872.

Clarke, M. L. *Paley: Evidences for the Man.* London: Society for Promoting Christian Knowledge, 1974.

"A Classbook of Botany." *American Journal of Science* 49 (1845): 190.

Clute, Willard N. "Botany Laboratory Equipment." *School Science and Mathematics* 18 (1918): 492–94.

Coburn, Louise H. *Kate Furbish, Botanist: An Appreciation.* N.p., 1924.

Cole, Arthur C. *A Hundred Years of Mount Holyoke College: The Evolution of an Educational Ideal.* New Haven, Conn.: Yale University Press, 1940.

"The Collection and Preservation of Plants." *Godey's Lady's Book* 66 (1863): 350–52.

Comstock, Anna Botsford. "The Evergreens." *Nature Study Review* 2 (1906): 7–13.

Comstock, J. L. *The Flora Belle: Or, Gems from Nature.* New York: T. L. Magagnos, 1847.

———. *An Introduction to the Study of Botany.* Hartford: D. F. Robinson and Co., 1832.

————. *The Young Botanist.* New York: Robinson, Pratt and Co., 1835.

Cooper, S[usan] F[enimore]. *Rural Hours.* New York: Putnam, 1850.

Coultas, Harland. "Hints to Ladies Studying Botany." *Godey's Lady's Book* 52 (1856): 514.

————. *The Principles of Botany, as Exemplified in the Cryptogamia.* Philadelphia: Lindsay and Blakiston, 1853.

————. *The Principles of Botany, as Exemplified in the Phanerogamia.* Philadelphia: King and Bards, 1854.

————. *What May Be Learned from a Tree.* New York: D. Appleton and Co., 1860.

————. "Wild Flowers." *Godey's Lady's Book* 44 (1852): 372–73, 485–86.

————. "The Wild Flowers of Early Spring Time." *Godey's Lady's Book* 48 (1854): 343–45.

Coulter, John M. "Botany as a Factor in Education." *School Review* 12 (1904): 609–17.

————. "Botany for High Schools and Colleges." *Botanical Gazette* 6 (1880): 96–98.

Cowen, David L. "Materia Medica and Pharmacology." In *The Education of American Physicians: Historical Essays*, edited by Ronald L. Numbers, 97–101. Berkeley: University of California Press, 1980.

Coxe, Margaret. *Floral Emblems: Or, Moral Sketches from Flowers.* Cincinnati: Henry W. Darby and Co., 1845.

Cravens, Hamilton. "American Science Comes of Age: An Institutional Perspective, 1830–1930." *American Studies* 17 (1976): 49–70.

Creevey, Caroline A. *Recreations in Botany.* New York: Harper, 1893.

"Curious Trees." *Youth's Cabinet* 1 (1846): 14–15.

Curti, Merle. *The Growth of American Thought.* 3d ed. New Brunswick, N.J.: Transition Books, 1982.

Curtis, A. H. "Hints on Herborizing." *American Naturalist* 6 (1872): 257–60.

Daley, Yvonne. "Vermont Fern Expert, Thriving at Age 93." *Boston Globe*, July 22, 1985, 17, 19.

Dana, Mrs. William Starr. *How to Know the Wild Flowers: A Guide to the Names, Haunts, Habits of Our Common Wild Flowers.* New York: Charles Scribner's Sons, 1899.

————. *Plants and Their Children.* New York: American Book Co., 1896.

Daniels, George H. *American Science in the Age of Jackson.* New York: Columbia University Press, 1968.

————. *Nineteenth-Century American Science: A Reappraisal.* Evanston, Ill.: Northwestern University Press, 1972.

————. "The Process of Professionalization in American Science: The Emergent Period, 1820–1860." *Isis* 58 (1967): 151–66.

————. *Science in American Society: A Social History.* New York: Knopf, 1971.

Darby, John. *Botany of the Southern States.* New York: A. S. Barnes, 1855.

————. *A Manual of Botany Adapted to the Productions of the Southern States.* Macon, Ga.: Benjamin F. Griffin, 1841.

Darlington, William. *Address to the Chester County Cabinet of Natural Sciences at the Organization of the Society.* West Chester, Pa.: The Cabinet, 1826.

Davidson, A. F. "Liberal Education for Farmers." *Country Gentleman* 16 (1860): 275.

Davis, Helen Burns. *Life and Work of Cyrus Guernsey Pringle.* Burlington, Vt.: University of Vermont Press, 1936.

————. "The Pringle Herbarium and Its Founder." *Vermont Alumnus* 17 (1938): 164–65.

Dean, Fanny A. *Little Talks about Plants: Easy Reading for Home and School.* Boston: D. Lothrop and Co., 1884.

Deane, Walter. "Thomas Morang." *Botanical Gazette* 19 (1894): 225–28.

"Death of an Old Botanist." *Botanical Gazette* 5 (1860): 150.

"DeCandolle's Botany." *North American Review* 38 (1834): 32–63.

Deiss, William A. "Spencer F. Baird and His Collectors." *Journal of the Society for the Bibliography of Natural History* 9 (1980): 635–45.

De Jong, John A. "American Attitudes toward Evolution before Darwin." Ph.D. diss., State University of Iowa, 1962.

Delta. "The Sea-Weed Album." *St. Nicholas* 2 (1874): 627–29.

Denner, Edith. "Science and Long Skirts." *Water Cure Journal* 20 (1855): 7.

Dennis, Mary B. *A Study of Leaves.* New York: D. Appleton and Co., 1888.

Dodge, Charles R[ichard]. "How to Prepare Sea-Mosses." *Field and Forest* 3 (1878): 168–70.

————. *Louise and I: A Seaside Story.* New York: G. W. Carleton and Co., 1879.

"Domestic Greenhouses." *Godey's Lady's Book* 20 (1840): 155–56.

Duncan, Florence I. *A Course of Lessons in Modeling Wax Flowers.* Philadelphia: J. B. Lippincott, 1887.

Dupree, A. Hunter. *Asa Gray, 1810–1888.* 1959. Reprint. New York: Atheneum, 1968.

————. *Science in the Federal Government: A History of Policies and Activities to 1940.* Cambridge, Mass.: Belknap Press of Harvard University Press, 1957.

E. F. B. "Botanizing at the Coal Bin: Examination of Specimens from the World's Large and Old Herbarium." *Observer* 1, no. 2 (February 1890): 1–2.

An Easy Introduction to the Knowledge of Nature. Philadelphia: American Sunday-School Union, 1846.

Eaton, Amos. *A Manual of Botany for the Northern and Middle States of America.* Albany, N.Y.: Websters and Skinners, 1824.

————. *A Manual of Botany for the Northern and Middle States of America.* Albany, N.Y.: O. Steele, 1833.

————. *A Manual of Botany for the Northern States.* Albany, N.Y.: Websters and Skinners, 1817.

Eaton, Amos, and John Wright. *North American Botany.* Troy, N.Y.: Elias Gates, 1840.

————. *North American Botany.* Troy, N.Y.: Elias Gates, 1887.

"The Economy and Habits of Plants." *Western Journal of Agriculture, Manufac-*

tures, Mechanic Arts, Internal Improvements, Commerce and General Literature 1 (1848): 10–25.

"Edible Fungi." *Appleton's Journal* 13 (1875): 604–5.

"Editors Table." *Godey's Lady's Book* 88 (1874): 469–70.

Egan, Maurice F. "A Bouquet of Roses." *Appleton's Journal* 6 (1871): 103–5.

Elliott, Clark A. "The American Scientist in Antebellum Society: A Quantitative View." *Social Studies of Science* 5 (1975): 93–108.

———. *Biographical Dictionary of American Science: The Seventeenth through the Nineteenth Centuries.* Westport, Conn.: Greenwood Press, 1979.

———. "Models of the American Scientist: A Look at Collective Biography." *Isis* 73 (1982): 77–102.

Elson, Ruth Miller. *Guardians of Tradition.* Lincoln: University of Nebraska Press, 1960.

Emerson, George B. "On the Education of Females." *Lectures of the American Institute of Instruction* 2 (1831): 15–41.

Emsby. "Systematic Botany Nevertheless." *Botanical Gazette* 6 (1881): 296–300.

Epis. Obs. "Young Naturalists." *Youth's Companion* 15 (1841–42): 135.

Estelle. "Botany, no. I: Flowers." *Youth's Companion* 25 (1851): 2.

Ewan, Joseph. *Rocky Mountain Naturalists.* Denver, Colo.: University of Denver Press, 1950.

———, ed. *A Short History of Botany in the United States.* New York: Hafner Publishing Co., 1969.

Fairbanks, Mrs. A. W., ed. *Emma Willard and Her Pupils, or, Fifty Years of Troy Female Seminary, 1822–1872.* New York: Mrs. Russell Sage, 1898.

Farber, Paul L. "The Transformation of Natural History in the Nineteenth Century." *Journal of the History of Biology* 15 (1982): 145–52.

Farlow, W. G. "The Task of American Botanists." *Popular Science Monthly* 31 (1887): 305–14.

Feith, Hugh G. "The Summer Vacation of a Party of Young Naturalists." *Tidings from Nature* 2 (1885): 26–29, 43–44, 58–59.

Fernald, Evelyn I. "Michael S. Bebb, Illinois Botanist and Letter-Writer." *Transactions of the Illinois Academy of Science* 34 (1941): 12–16.

Fernald, M. L. "Some Early Botanists of the American Philosophical Society." *Proceedings of the American Philosophical Society* 86 (1942): 63–71.

Filler, Louis, ed. *An Ohio Schoolmistress: The Memoirs of Irene Hardy.* Kent, Ohio: Kent State University Press, 1980.

Flagg, Wilson. "Botanizing." *Atlantic Monthly* 27 (1871): 657–64.

———. *Studies in the Field and Forest.* Boston: Little, Brown and Co., 1857.

"Florida Plants." *Bulletin of the Torrey Botanical Club* 7 (1880): n.p.

"Flowers." *Merry's Museum* 8 (1844): 162.

Ford, Charles E. "Botany Texts: A Survey of Their Development in American Higher Education, 1643–1906." *History of Education Quarterly* 4 (1964): 59–71.

Ford, Charlotte A. "Eliza Frances Andrews, Practical Botanist." *Georgia Historical Quarterly* 70 (1986): 63.

"Four 'Inland' Girls by the Sea." *St. Nicholas* 5 (1877–78): 763–64.

Freidson, Eliot. *Professional Powers: A Study of the Institutionalization of Formal Knowledge.* Chicago: University of Chicago Press, 1986.

Fuller, Jane Jay. *Uncle John's Flower Gatherers: A Companion for the Woods and Fields.* New York: M. W. Dodd, 1869.

G. B. H. "A Little Botanical Knowledge." *Country Gentleman* 15 (1860): 47.

Geary, Sheila Connor, and B. June Hutchinson. "Mr. Dawson, Plantsman." *Arnoldia* 40 (1980): 51–75.

Gee, Wilson. *South Carolina Botanists: Biography and Bibliography.* University of South Carolina Bulletin no. 72. Columbia: University of South Carolina Press, 1918.

Geiser, Samuel Wood. *Naturalists on the Frontier.* Dallas: Southern Methodist University Press, 1937.

Geison, Gerald L. *Professions and Professional Ideologies in America.* Chapel Hill: University of North Carolina Press, 1983.

Gerstner, Patsy A. "The Academy of Natural Sciences of Philadelphia." In *The Pursuit of Knowledge in the Early American Republic: American Scientific and Learned Societies from Colonial Times to the Civil War*, edited by Alexandra Oleson and Sanborn C. Brown, 174–93. Baltimore, Md.: Johns Hopkins University Press, 1976.

Gibbons, Felton, and Deborah Strom. *Neighbors to the Birds: A History of Birdwatching in America.* New York: W. W. Norton, 1988.

Gilbert, B. D. "Historical Sketch of the Linnaean Fern Chapter." *Fern Bulletin* 10 (1902): 116–20.

Gillespie, Neal C. "Preparing for Darwin: Conchology and Natural Theology in Anglo-American Natural History." *Studies in History of Biology* 7 (1984): 93–145.

Gillman, Henry. "Our Northern Orchids." *Appleton's Journal* 9 (1873): 431.

The Girls' Manual: Comprising a Summary View of Female Studies, Accomplishments and Principles of Conduct. New York: D. Appleton and Co., 1865.

Glick, Wendell. "Bishop Paley in America." *New England Quarterly* 27 (1954): 347–54.

Goldstein, Jan. "Foucault among the Sociologists: The 'Disciplines' and the History of the Professions." *History and Theory* 23 (1984): 170–92.

Goodale, George L. *Concerning a Few Common Plants.* Boston Society of Natural History, Guides for Science Teaching, no. 2. Boston: Boston Society of Natural History, 1879.

———. "The Development of Botany as Shown in this Journal." *American Journal of Science*, 4th ser., 46 (1918): 399–416.

[Goodrich, Samuel G.]. "About the Leaves of Trees." *Parley's Magazine* 3 (1835): 41–42.

———. *Parley's Panorama: or, Curiosities of Nature and Art, History & Biography.* Hartford, Conn.: House and Brown, 1851.

Goodsell, Willystine, ed. *Pioneers of Women's Education in the United States: Emma Willard, Catharine Beecher, and Mary Lyon.* New York: McGraw Hill Publishing Co., 1931.

Gould, A. A. "On the Introduction of Natural History as a Study to Common Schools." *Lectures of the American Institute of Instruction* 5 (1834): 227–45.

Graustein, Jeannette E. *Thomas Nuttall, Naturalist: Explorations in America, 1808–1841.* Cambridge, Mass.: Harvard University Press, 1967.

Gray, Asa. *The Botanical Text-Book: An Introduction to Scientific Botany.* New York: George P. Putnam and Co., 1853.

———. *The Botanical Text-Book for Colleges, Schools and Private Students.* New York: Wiley and Putnam, 1842.

———. *Botany for Young People and Common Schools. Part I: How Plants Grow.* New York: Ivison, Blakeman, Taylor, and Co., 1858.

———. *Botany for Young People. Part II: How Plants Behave.* New York: Ivison, Blakeman, Taylor, and Co., 1872.

———. *Darwiniana: Essays and Reviews Pertaining to Darwinism.* Cambridge, Mass.: Harvard University Press, 1963.

———. *Elements of Botany.* New York: G. and C. Carvill and Co., 1836.

———. *The Elements of Botany: For Beginners and Schools.* New York: American Book Co., 1887.

———. *First Lessons in Botany and Vegetable Physiology.* New York: Ivison and Phinney and G. P. Putnam and Co., 1857.

———. *Letters of Asa Gray.* Edited by Jane Loring Gray. 2 vols. Boston: Houghton, Mifflin and Co., 1893.

———. "On the Importance of the Natural Sciences in Our System of Popular Education." *Lectures of the American Institute of Instruction* 12 (1841): 95–117.

Gray, Asa, and John Torrey. *Flora of North America.* New York: Wiley and Putnam, 1838–43.

"Gray Memorial Botanical Chapter of the Agassiz Association: Reports—Second Quarter, 1892." *Observer* 3 (1892): 245–46.

"Gray's *Manual of Botany* and *Flora of North America.*" *North American Review* 67 (1848): 174–93.

Green, Elizabeth Alden. *Mary Lyon and Mount Holyoke: Opening the Gates.* Hanover, N.H.: University Press of New England, 1979.

Green, Frances H., and J. W. Congdon. *Analytical Class-Book of Botany.* New York: D. Appleton and Co., 1855.

Green, Harvey. *Fit for America: Health, Fitness, Sport, and American Society.* New York: Pantheon, 1986.

Greene, John C. *American Science in the Age of Jefferson.* Ames: Iowa State Press, 1984.

———. "Science and Religion." In *The Rise of Adventism: Religion and Society in Mid-Nineteenth-Century America,* edited by Edwin S. Gaustad, 50–69. New York: Harper and Row, 1974.

———. "Science and the Public in the Age of Jefferson." *Isis* 49 (1958): 13–25.

Greene, Welcome Arnold, et al. *Providence Plantations for Two Hundred and Fifty Years.* Providence, R.I.: J. A. and R. A. Reid, 1886.

Guralnick, Stanley M. "Geology and Religion before Darwin: The Case of Edward Hitchcock, Theologian and Geologist (1793–1864)." *Isis* 63 (1972): 529–43.

———. *Science and the Ante-Bellum American Colleges.* Memoirs, vol. 109. Philadelphia: American Philosophical Society, 1975.

———. "Sources of Misconception on the Role of Science in the Nineteenth-Century American College." *Isis* 65 (1974): 352–66.

Hall, Abbie G. *Botany : Lessons in Botany and Analysis of Plants.* Chicago: Geo. Sherwood and Co., 1887.

Haller, John S., Jr. *American Medicine in Transition, 1840–1910.* Urbana: University of Illinois Press, 1981.

Halsted, Byron D. "What the Station Botanists Are Doing." *Botanical Gazette* 16 (1891): 288–91.

Hardy, Stephen. *How Boston Played: Sport, Recreation, and Community, 1865–1915.* Boston: Northeastern University Press, 1982.

Harrington, Mark W. *The Analysis of Plants Intended for Schools and Colleges and for the Independent Botanical Student.* Ann Arbor, Mich.: Sheehan and Co., 1880.

Harris, Amanda B. *Wild Flowers and Where They Grow.* Boston: D. Lothrop and Co., 1882.

Harshberger, John W. *The Botanists of Philadelphia and Their Work.* Philadelphia: T. C. Davis and Son, 1899.

Haskell, Thomas L., ed. *The Authority of Experts.* Bloomington: Indiana University Press, 1984.

———. *The Emergence of Professional Social Science: The American Social Science Association and the Nineteenth-Century Crisis of Authority.* Chicago: University of Illinois Press, 1977.

Hawkes, Graham Parker. "Increase A. Lapham: Wisconsin's First Scientist." Ph.D. diss., University of Wisconsin, 1960.

Hawks, Francis Lister. *The American Forest: Or, Uncle Philip's Conversations with the Children about the Trees of America.* New York: Harper and Brothers, 1834.

Haygood, Tamara Minor. *Henry William Ravenel, 1814–1887: South Carolina Scientist in the Civil War Era.* Tuscaloosa: University of Alabama Press, 1987.

———. "Spheres of Influence in American Botanical Leadership in the Nineteenth Century." Paper delivered at the annual meeting of the History of Science Society, Los Angeles, 29 December 1981.

Hayner, Rutherford. *Troy and Rensselaer County, New York: A History.* New York: Lewis Historical Publishing Co., 1925.

Hazzard, W. W. "On the Study of Plants." *Southern Agriculturist* 3 (1830): 7–11.

Hendrickson, Walter. *The Arkites, and Other Pioneer Natural History Organizations of Cleveland.* Makers of Cleveland Series, no. 1. Cleveland: Press of Western Reserve University, 1962.

———. "Science and Culture in Nineteenth-Century Michigan." *Michigan History* 57 (1973): 140–50.

————. "Science and Culture in the American Middle West." *Isis* 64 (1973): 326–40.

Hersh, Blanche Glassman. "The 'True Woman' and the 'New Woman' in Nineteenth-Century America: Feminist-Abolitionists and a New Concept of True Womanhood." In *Woman's Being, Woman's Place: Female Identity and Vocation in American History*, edited by Mary Kelley, 271–82. Boston: G. K. Hill and Co., 1979.

Hervey, A. B. *Sea-Mosses: A Collector's Guide and Introduction to the Study of Marine Algae.* Boston: S. E. Cassino, 1881.

Higgins, John. "Vestiges." *Observer* 4 (1893): 153–54.

[Higginson, Thomas Wentworth]. "The Health of Our Girls." *Atlantic Monthly* 9 (1862): 722–31.

Higham, John. "The Matrix of Specialization." In *The Organization of Knowledge in Modern America, 1860–1920*, edited by Alexandra Oleson and John Voss, 3–18. Baltimore, Md.: Johns Hopkins University Press, 1979.

Hindle, Brooke. *The Pursuit of Science in Revolutionary America.* 1956. Reprint. New York: W. W. Norton and Co., 1974.

Hovenkamp, Herbert. *Science and Religion in America, 1800–1860.* Philadelphia: University of Pennsylvania Press, 1978.

"How to Converse." *Work and Play* 1 (1875): 94–96.

Humphrey, Harry Baker. *Makers of North American Botany.* New York: Ronald Press Co., 1901.

I. C. "Divinity of Nature." *Godey's Lady's Book* 50 (1855): 208.

Inglis, Alexander James. *The Rise of the High School in Massachusetts.* New York: Teachers College, 1911.

Irving, C. *A Catechism of Botany.* New York: Collins and Hannay, 1829.

James, Mrs. I. "Gray's Botanical Text-Books." *North American Review* 87 (1858): 321–42.

Johnson, Laura. *Botanical Teacher for North America.* Albany, N.Y.: Oliver Steele, 1834.

Johnson, Thomas Cary. *Scientific Interest in the Old South.* New York: D. Appleton-Century Co., 1936.

Johnson, Thomas H. *The Complete Poems of Emily Dickinson.* London: Faber and Faber, 1970.

Kaestle, Carl. "The History of Literacy and the History of Readers." *Review of Research in Education* 12 (1985): 11–53.

Kastner, Joseph. *A Species of Eternity.* New York: Alfred A. Knopf, 1977.

————. *A World of Watchers.* New York: Alfred A. Knopf, 1986.

Katz, Michael B. *The Irony of Early School Reform: Educational Innovation in Mid-Nineteenth-Century Massachusetts.* Boston: Beacon Press, 1968.

Kaufman, Martin. "American Medical Education." In *The Education of American Physicians: Historical Essays*, edited by Ronald L. Numbers, 7–28. Berkeley: University of California Press, 1980.

Kelley, Howard Atwood. *Some American Medical Botanists.* New York: Appleton, 1929.

Kiefer, Monica Mary. *American Children through Their Books, 1700–1835.* Philadelphia: University of Pennsylvania Press, 1948.

King, F. H. *A Scheme for Thorough, Rapid, Systematic Plant Analysis.* Berlin, Wis.: Courant Offices, 1876.

Kirkwood, J. E. "Opportunity and Obligation in Botanical Teaching." *School and Science Mathematics* 18 (1918): 579–87.

Knight, Mrs. A. A. *A Primer of Botany.* Boston: Ginn and Co., 1887.

Knowlton, D. H. *Plant Study for Children.* Farmington, Me.: D. H. Knowlton and Co., 1882.

Kohlstedt, Sally Gregory. *The Formation of the American Scientific Community: The American Association for the Advancement of Science, 1848–60.* Urbana: University of Illinois Press, 1976.

———. "The Nineteenth-Century Amateur Tradition: The Case of the Boston Society of Natural History." In *Science and Its Public: The Changing Relationship,* edited by Gerald Holton and William A. Blanpied, 173–90. Dordrecht, Holland: D. Reidel Co., 1976.

———. "Parlors, Primers, and Public Schooling: Education for Science in Nineteenth-Century America." *Isis* 81 (1990), 425–45.

Krug, Edward A. *The Shaping of the American High School, 1880–1920.* Madison: University of Wisconsin Press, 1969.

Kuritz, Hyman. "The Popularization of Science in Nineteenth-Century America." *History of Education Quarterly* 21 (1981): 259–74.

L. "Threads from the Web of Aunt Kate's Life." *Youth's Cabinet* 6 (1851): 315–18.

Laura. "Letter from a Correspondent." *Merry's Museum* 3 (1842): 89–90.

LeMahieu, D. L. *The Mind of William Paley: A Philosopher and His Age.* Lincoln: University of Nebraska Press, 1976.

"A Lesson in Botany." *American Journal of Education* 4 (1829): 254–56.

"Lessons in Botany." *Parley's Magazine* 3 (1835): 66–68.

"Lessons in Botany—No. II." *Parley's Magazine* 3 (1835): 90–91.

"Lessons Whispered on a Walk." *Country Gentleman* 1 (1853): 168–69.

"A Letter from the School Girls." *Youth's Cabinet* 2 (1847): 164–65.

"Letter of P. R. Hoy." In the *Annual Report of the State Superintendent of Public Instruction, of the State of Wisconsin for the Year 1854,* 78–79. Madison, Wis.: Beriah Brown, 1855.

Leverette, William Edward, Jr. "Science and Values: A Study of Edward L. Youmans' *Popular Science Monthly,* 1872–1887." Ph.D. diss., Vanderbilt University, 1963.

"Lichens." *Harper's New Monthly Magazine* 15 (1857): 177–83.

Lincoln Phelps, Almira Hart. *Botany for Beginners.* Hartford, Conn.: F. J. Huntington, 1833.

———. *Botany for Beginners.* Philadelphia: J. B. Lippincott Co., 1836.

———. *Botany for Beginners.* New York: F. J. Huntington and Co., 1837.

———. *Botany for Beginners.* Philadelphia: J. B. Lippincott Co., 1891.

———. *Caroline Westerley: or, The Young Traveller from Ohio.* New York: J. and J. Harper, 1833.

———. *The Educator: or, Hours with My Pupils.* New York: A. S. Barnes, 1876.

———. *Familiar Lectures on Botany.* Hartford, Conn.: H. and F. J. Huntington, 1829.

———. *Familiar Lectures on Botany.* Hartford, Conn.: H. and F. J. Huntington, 1831.

———. *Familiar Lectures on Botany.* Hartford, Conn.: F. J. Huntington, 1836.

———. *Familiar Lectures on Botany.* New York: Francis J. Huntington, 1851.

———. *Familiar Lectures on Botany.* New York: Francis J. Huntington, 1852.

———. *The Fireside Friend, or Female Student: Being Advice to Young Ladies on the Important Subject of Education.* Boston: Marsh, Capen, Lyon and Webb, 1840.

———. *The Fireside Friend, or Female Student: Being Advice to Young Ladies on the Important Subject of Education.* New York: Harper, 1855.

———. *Lectures to Young Ladies: Comprising Outlines and Applications of the Different Branches of Female Education.* Boston: Carter, Hendee and Co., 1833.

[———.] "Popular Botany." *National Quarterly Review* 2 (1860–61): 276–96.

Locke, John. *Outlines of Botany.* Boston: Cummings and Hilliard, 1819.

Lorimer, Mary. *Among the Trees: A Journal of Walks in the Woods and Flower-Hunting through Field and by Brook.* New York: Hurd and Houghton, 1869.

Lounsberry, Alice. *A Guide to the Wild Flowers.* New York: Frederick A. Stokes Co., 1889.

Loveland, Mary A. "A Half-Mile Walk." *Observer* 4 (1893): 373–74.

Lutz, Alma. *Emma Willard: Daughter of Democracy.* Boston: Houghton Mifflin Co., 1929.

Lyon, Charles J. "Centennial of Wood's 'Class-Book of Botany.'" *Science* 101 (1945): 484–86.

M. D. R. "A Summer Day's Pastime." *Our Young Folks* 6 (1870): 502–6.

M. E. B. "Notes on the Fungi of Maryland." *Field and Forest* 3 (1877–78): 63.

M. W. D. "Spring Flowers." *Youth's Companion* 30 (1856): 8.

McAllister, Ethel. *Amos Eaton: Scientist and Educator.* Philadelphia: University of Pennsylvania Press, 1941.

McAtee, W. L., ed. "Journal of Benjamin Smith Barton on a Visit to Virginia, 1802." *Castanea* 3 (1938): 85–117.

McDonald, Wm. H. "Autobiographical Notes." *Asa Gray Bulletin* 5 (1897): 22–24.

McGuffey, W. H. *McGuffey's New Fourth Eclectic Reader.* Cincinnati, Ohio: Van Antwerp, Bragg and Co., 1866.

McKelvey, Susan D. *Botanical Exploration of the Trans-Mississippi West, 1790–1850.* Jamaica Plain, Mass.: Arnold Arboretum, 1956.

Macloskie, George. *Elementary Botany with Student Guide to the Examination and Description of Plants.* New York: Henry Holt and Co., 1883.

"Maria's Visit." *Juvenile Miscellany* 5 (1833–34): 33–62.

Mavor, William. *Catechism of Botany: or, An Easy Introduction to the Vegetable Kingdom.* Boston: J. Belcher, 1811.

Mayer, William V. "Biology Education in the United States during the Twentieth Century." *Quarterly Review of Biology* 61 (1986): 482–84.

Meadows, Jack, and Tim Fisher. "Gentlemen v. Players." *New Scientist* 79 (1978): 752–54.

Meisel, Max. *A Bibliography of American Natural History: The Pioneer Century, 1769–1865.* 3 vols. 1924. Facsimile reprint. New York: Hafner Publishing Co., 1967.

Melder, Keith. "The Mask of Oppression: The Female Seminary Movement in the United States." *New York History* 55 (1974): 261–79.

Miller, Eli B. *The Botanist's Plant Record, for High Schools, Academies, & Colleges.* Battle Creek, Mich.: Review and Herald Press, 1882.

Miller, George Frederick. *The Academy System of the State of New York.* Albany, N.Y.: J. B. Lyon, 1922.

Miller, Howard S. *Dollars for Research: Science and Its Patrons in Nineteenth-Century America.* Seattle: University of Washington Press, 1970.

Minton, Tyree G. "The History of the Nature-Study Movement and Its Role in the Development of Environmental Education." Ph.D. diss., University of Massachusetts, 1980.

Mitchell, S. W. "Lectures on the Conduct of the Medical Life." *University Medical Magazine* 5 (1893): 651–74.

"The Moral Uses of Plants." *Western Journal and Civilian* 1 (1848): 39–46.

Mott, Frank Luther. *A History of American Magazines.* 5 vols. Cambridge, Mass.: Harvard University Press, 1938–68.

"Mrs. Almira Lincoln Phelps." *American Journal of Education* 1 (1867): 611–20.

"The Mystery of Nature." *Atlantic Monthly* 20 (1867): 349.

The National Union Catalogue: Pre-1956 Imprints. London: Mansell, 1971.

Newbury, A. "Weeds and Words." *Our Young Folks* 7 (1871): 108–9.

Newell, Jane H. *Outlines of Lessons in Botany for the Use of Teachers, or Mothers Studying with Their Children.* Boston: Ginn and Co., 1897.

Newman, John B. *Boudoir Botany: or, The Parlor Book of Flowers.* New York: Harper and Brothers, 1847.

———. *The Illustrated Botany.* New York: J. K. Wellman, 1846.

Noll, Henry R. *The Botanical Class-Book and Flora of Pennsylvania.* Lewisburg, Pa.: O. N. Worden, 1852.

"Notice of the Late Dr. Waterhouse." *Journal of the Philadelphia Academy of Natural Sciences* 1 (1817–18): 31–32.

Numbers, Ronald. *Creation by Natural Law: Laplace's Nebular Hypothesis in American Thought.* Seattle: University of Washington Press, 1977.

Nye, Russel. *The Unembarrassed Muse: The Popular Arts in America.* New York: Dial Press, 1970.

Oleson, Alexandra, and Sanborn C. Brown, eds. *The Pursuit of Knowledge in the Early American Republic: American Scientific and Learned Societies from Colonial Times to the Civil War.* Baltimore, Md.: Johns Hopkins University Press, 1976.

Oleson, Alexandra, and John Voss, eds. *The Organization of Knowledge in*

Modern America, 1860–1920. Baltimore, Md.: Johns Hopkins University Press, 1979.

"On Botany." *Southern Literary Messenger* 7 (1841): 777–81.

"On Plants." *Youth's Companion* 9 (1835–36): 35.

"Our Native Orchids." *Appleton's Journal* 5 (1871): 706–7.

Overfield, Richard A. "Charles E. Bessey: The Impact of the 'New' Botany on Agriculture, 1880–1910." *Technology and Culture* 16 (1975): 162–81.

Owen, M. L. "Botany in Schools." *Old and New* 6 (1872): 245–48.

Palmer, Julius A. "Toadstool-Eating." *Popular Science Monthly* 11 (1877): 93–100.

Parry, C. C. "Herbarium Cases." *American Naturalist* 8 (1873): 471–73.

The Pastime of Learning with Sketches of Rural Scenes. Boston: Cottons and Barnard, 1831.

Pauly, Philip J. "The Appearance of Academic Biology in Late Nineteenth-Century America." *Journal of the History of Biology* 17 (1984): 369–97.

Perry, Lewis. *Intellectual Life in America: A History*. New York: Franklin Watts, 1984.

Phelps, Almira Hart Lincoln. *See* Lincoln Phelps, Almira Hart.

Phelps, Elizabeth Stuart. *Doctor Zay*. 1882. Reprint. New York: Feminist Press at the City University of New York, 1987.

Phoebus, Mrs. V. C. *Plant Life*. Home College Series, no. 53. New York: Phillips and Hunt, 1883.

"Plants for Sale." *Botanical Gazette* 2 (1877): 147.

Plimpton, Pauline Ames, ed. *Oakes Ames: Jottings of a Harvard Botanist, 1874–1950*. Cambridge, Mass.: Botanical Museum of Harvard University, 1979.

Porter, Roy. "Gentlemen and Geology: The Emergence of a Scientific Career, 1660–1920." *Historical Journal* 21 (1978): 809–36.

R. E. "A Talk about Trees." *Youth's Companion* 12 (1838–39): 42–43.

Ragland, Fannie. "Weed Study in Grammar Grades." *Nature Study Review* 7 (1911): 167–70.

Ramsay, David. *History of South Carolina*. 2 vols. South Carolina Heritage Series, no. 4. Newberry, S.C.: W. W. Duffie, 1960.

Reed, Howard S. *A Short History of the Plant Sciences*. Waltham, Mass.: Chronica Botanica Co., 1942.

Reifschneider, Olga. *Biographies of Nevada Botanists, 1844–1963*. Reno: University of Nevada Press, 1964.

Reingold, Nathan. "American Indifference to Basic Research: A Reappraisal." In *Nineteenth-Century American Science: A Reappraisal*, edited by George H. Daniels, 38–62. Evanston, Ill.: Northwestern University Press, 1972.

———. "Definitions and Speculations: The Professionalization of Science in America in the Nineteenth Century." In *The Pursuit of Knowledge in the Early American Republic: American Scientific and Learned Societies from Colonial Times to the Civil War*, edited by Alexandra Oleson and Sanborn C. Brown, 33–69. Baltimore, Md.: Johns Hopkins University Press, 1976.

Rennie, James, and Arabella Clark. *An Alphabet of Botany for the Use of Beginners*. New York: Peter Hill, 1833.

Reverchon, J. "Botanizing in Texas." *Botanical Gazette* 11 (1886): 56–59, 211–16.

"Review of Lincoln's *Familiar Lectures on Botany*." *National Quarterly Review* 1 (1860): 531–32.

Reznick, Samuel. *Education for a Technological Society: A Sesquicentennial History of Rensselaer Polytechnic Institute*. Troy, N.Y.: Rensselaer Polytechnic Institute, 1968.

Rickett, H. W., ed. *Botanic Manuscript of Jane Colden, 1724–1766*. New York: Garden Club of Orange and Dutchess Counties, 1963.

Riddell, J. L. "Brief Sketch of Subjects Embraced in the Science of Botany with Its Relation to Medicine, and Some of the Inducements for Engaging in Its Study." *New Orleans Medical and Surgical Journal* 11 (1846): 445–49.

Riley, Glenda. *Inventing the American Woman: A Perspective on Woman's History, 1607–1877*. Arlington Heights, Ill.: Harlan Davidson, 1986.

Robinson, John. *Ferns in Their Homes and Ours*. Salem, Mass.: S. E. Cassino, 1878.

"Rock, Tree, and Man." *Atlantic Monthly* 4 (1859): 29–40.

Rodgers, Andrew Denny, III. *American Botany, 1873–1892: Decades of Transition*. 1944. Facsimile reprint. New York: Hafner Publishing Co., 1968.

———. *John Merle Coulter: Missionary in Science*. Princeton, N.J.: Princeton University Press, 1944.

———. *John Torrey: A Story of North American Botany*. Princeton, N.J.: Princeton University Press, 1942.

———. *Liberty Hyde Bailey: A Story of American Plant Sciences*. 1949. Facsimile reprint. New York: Hafner Publishing Co., 1965.

Rodgers, Daniel T. *The Work Ethic in Industrial America, 1850–1920*. Chicago: University of Chicago Press, 1978.

Roessler, S. E. "Outdoors." *Observer* 1, no. 5 (May 1890): 3.

———. "Summering." *Observer* 1, no. 7 (July 1890): 2.

Roorbach, O. A. *Bibliotheca Americana: Catalogue of American Publications Including Reprints and Original Works from 1820 to 1852*. New York: privately printed, 1852.

Ross, Sydney. "*Scientist*: The Story of a Word." *Annals of Science* 18 (1962): 65–85.

Rossiter, Margaret W. *Women Scientists in America: Struggles and Strategies to 1940*. Baltimore, Md.: Johns Hopkins University Press, 1982.

Rothenberg, Marc. "Organization and Control: Professionals and Amateurs in American Astronomy, 1899–1918." *Social Studies of Science* 11 (1981): 305–25.

Rudolph, Emanuel D. "Almira Hart Lincoln Phelps." *American Journal of Botany* 71 (1984): 1161–67.

———. "How It Developed That Botany Was the Science Thought Most Suitable for Victorian Young Ladies." *Children's Literature* 2 (1973): 92–97.

———. "The Introduction of the Natural System of Classification of Plants to Nineteenth-Century American Students." *Archives of Natural History* 10 (1982): 461–68.

————. "Women in Nineteenth-Century American Botany: A Generally Un-
recognized Constituency." *American Journal of Botany* 69 (1982): 1346–55.

Russell, John L. "Mushrooms." *American Naturalist* 2 (1868–69): 292–303.

Sandburg, Carl. *Always the Young Strangers.* New York: Harcourt, Brace and
Co., 1952.

Sanders, Charles W. *The New School Reader: Fourth Book.* New York: Ivison
and Phinney, 1857.

————. *Sanders' Union Reader: Number Three.* New York: Ivison, Phinney,
Blakeman and Co., 1862.

Savage, Henry, Jr. *Discovering America, 1700–1875.* New York: Harper and
Row, 1979.

Schmitt, Peter J. *Back to Nature: The Arcadian Myth in Urban America.* New
York: Oxford University Press, 1969.

Scott, Anne Firor. "Almira Hart Lincoln Phelps: The Self-Made Woman in the
Nineteenth Century." *Maryland Historical Magazine* 75 (1980): 203–16.

————. "The Ever Widening Circle: The Diffusion of Feminist Values from
the Troy Female Seminary: 1822–1872." *History of Education Quarterly* 19
(1979): 3–25.

————. *Making the Invisible Woman Visible.* Urbana: University of Illinois
Press, 1984.

Setchell, William Albert. "Townshend Stith Brandegee and Mary Katherine
(Layne) (Curran) Brandegee." *University of California Publications in Botany*
13 (1924–27): 165–78.

Shecut, J. L. E. W. *Medical and Philosophical Essays.* Charleston, S.C.: privately printed, 1819.

Shepard, C. A. "Botany." *Observer* 2, no. 5 (May 1891): 4.

————. "Botany." *Observer* 4 (1893): 85.

————. "Botany's Charms." *Observer* 2, no. 1 (January 1891): 5.

Sherwin, Thomas. "The Relative Importance of Ancient Classical and Scientific Studies." *Massachusetts Teacher* 9 (1856): 398–404.

Shils, Edward. "The Order of Learning in the United States: The Ascendancy
of the University." In *The Organization of Knowledge*, edited by Alexandra
Oleson and John Voss, 19–47. Baltimore, Md.: Johns Hopkins University
Press, 1979.

Shteir, Ann B. "Botany in the Breakfast Room: Women and Early 19th-
Century British Plant Study." In *Uneasy Careers and Intimate Lives: Women
in Science, 1789–1979*, edited by Pnina Abir-am and Dorinda Outram, 31–
43. New Brunswick, N.J.: Rutgers University Press, 1982.

————. "Linnaeus's Daughters: Women and British Botany." In *Women and
the Structure of Society: Selected Research from the Fifth Berkshire Conference on History of Women*, edited by Barbara J. Harris and JoAnn K.
McNamara, 67–73, 261–63. Durham, N.C.: Duke University Press, 1984.

Silber, Kate. *Pestalozzi: The Man and His Work.* New York: Schocken Books,
1973.

Sizer, Theodore R. *Secondary Schools at the Turn of the Century.* New Haven,
Conn.: Yale University Press, 1964.

"Sketch of Prof. Gray." *Popular Science Monthly* 1 (1872): 490–95.

Sklar, Katheryn Kish. *Catharine Beecher: A Study in American Domesticity.* New York: W. W. Norton and Co., 1976.

Sloan, Douglas. "Science in New York City, 1867–1907." *Isis* 71 (1980): 35–76.

Smallwood, William Martin. "Amos Eaton, Naturalist." *New York History* 18 (1937): 167–88.

Smallwood, William Martin, and Mabel Sarah Coon Smallwood. *Natural History and the American Mind.* New York: Columbia University Press, 1941.

Smith, Elias. *The American Physician and Family Assistant.* Boston: E. Bellamy, 1826.

Smith, Mrs. Harrison. "The Season of Flowers." *Godey's Lady's Book* 15 (1837): 12.

Smith-Rosenberg, Carroll. *Disorderly Conduct: Visions of Gender in Victorian America.* New York: Oxford University Press, 1985.

A Southerner. "Life in the Southwest." *Youth's Cabinet* 2 (1847): 251–52.

Spaulding, V. M. "Botany in the High School." *Academy* 5 (1891): 313–21.

Spears, Betty, and Richard A. Swanson. *History of Sport and Physical Activity in the United States.* Dubuque, Iowa: Wm. C. Brown, 1983.

Stearns, Raymond P. *Science in the British Colonies of America.* Urbana: University of Illinois Press, 1970.

Stebbins, Robert A. "The Amateur: Two Sociological Definitions." *Pacific Sociological Review* 20 (1977): 582–606.

———. *Amateurs: On the Margin between Work and Leisure.* Beverly Hills, Calif.: Sage Publications, 1979.

Steere, W. C., ed. *Fifty Years of Botany: Golden Jubilee Volume of the Botanical Society of America.* New York: McGraw-Hill, 1958.

Stout, John Elbert. *The Development of High School Curriculum.* New York: Harper and Brothers, 1960.

Stuckey, Ronald L., ed. *The Development of Botany in Selected Regions of North America before 1900.* New York: Arno Press, 1978.

"The Study of Botany in Common Schools." *Country Gentleman* 3 (1854): 189.

"The Study of Flowers." *Merry's Museum* 8 (1847): 158–59.

Tarver, Micajah. *The Moral Uses of Plants.* St. Louis, Mo.: C. Witter, 1855.

"Testimony of George B. Emerson to the Committee on Education, Massachusetts State Legislature; re. Subsidizing Distribution of *The American Naturalist* for Teachers." *American Naturalist* 5 (1870–71): 132–34.

Thinker, Theodore. "The Early Spring Flowers." *Youth's Cabinet* 4 (1846): 102.

———. *First Lessons in Botany.* New York: A. S. Barnes and Co., 1851.

———. "The Transplanted Tree." *Youth's Cabinet* 2 (1847): 10.

Thomas, John L. "Romantic Reform in America, 1815–1865." In *Ante-Bellum Reform*, edited by David Brion Davis, 153–76. New York: Harper and Row, 1967.

Thomson, Samuel. *New Guide to Health or Botanic Family Physician.* Boston: privately printed, 1822.

Thoreau, Henry D. *The Journal of Henry D. Thoreau.* Edited by Bradford Torrey and Francis H. Allen. Rev. ed. in 2 vols. New York: Dover Publications, 1962.

Tilgate. "Botany." *Country Gentleman* 15 (1856): 400.

Tippo, Oswald. "The Early History of the Botanical Society of America." In *Fifty Years of Botany: Golden Jubilee Volume of the Botanical Society of America,* edited by William Campbell Steere, 1–13. New York: McGraw-Hill, 1958.

"To the Invisible *Author* of Nature.—Selected." *Christian Disciple* 3 (1815): 320.

Tobey, Ronald C. *Saving the Prairies: The Life Cycle of the Founding School of American Plant Ecology, 1895–1955.* Berkeley: University of California Press, 1981.

Treat, Mary. "Botany for Invalids." *Herald of Health* 42 (1866): 39–40, 70–72, 125–27, 164–65.

"Trees in Winter." *Observer* 1, no. 2 (February 1890): 2.

True, Alfred Charles. *A History of Agricultural Education in the United States, 1785–1925.* 1929. Reprint. New York: Arno Press, 1980.

Turner, Cordelia Harris. *The Floral Kingdom: Its History, Sentiment and Poetry.* Chicago: Standard-Columbian Co., 1891.

Turner, Nancy Byrd. *Testament of Happiness: Letters of Annie Oakes Huntington.* Portland, Me.: Anthoeson Press, 1947.

Underhill, Orra E. *The Origins and Development of Elementary-School Science.* Chicago: Scott, Foresman and Co., 1941.

Underwood, Lucien M. "The Study of Botany in High School." *Journal of Pedagogy* 11 (1898): 181–91.

Verbrugge, Martha H. *Able-Bodied Womanhood: Personal Health and Social Change in Nineteenth-Century Boston.* New York: Oxford University Press, 1988.

Veysey, Laurence, ed. *The Perfectionists: Radical Social Thought in the North, 1815–1860.* New York: John Wiley and Sons, 1973.

———. "Who's a Professional? Who Cares?" *Reviews in American History* 3 (1975): 419–23.

Voss, Edward Groesbeck. *Botanical Beachcombers and Explorers: Pioneers of the Nineteenth Century on the Upper Great Lakes.* Contributions from the University of Michigan Herbarium, vol. 13. Ann Arbor: University of Michigan Herbarium, 1978.

Wakefield, Priscilla. *An Introduction to Botany.* 6th ed. Philadelphia: Kimber and Conrad, 1811.

"The Walking-Fern." *Appleton's Journal,* n.s., 2 (1877): 442–53.

Walsh, Thomas R. "Charles E. Bessey and the Transformation of the Industrial College." *Nebraska History* 52 (1971): 383–409.

Walters, Ronald G. *American Reformers, 1815–1860.* American Century Series. New York: Hill and Wang, 1978.

Warner, Deborah Jean. *Graceanna Lewis: Scientist and Humanitarian.* Washington, D.C.: Smithsonian Institution Press, 1979.

———. "Science Education for Women in Ante-bellum America." *Isis* 69 (1978): 58–67.

Warner, John Harley. "'Exploring the Inner Labyrinths of Creation': Popular Microscopy in Nineteenth-Century America." *Journal of the History of Medicine and Allied Sciences* 37 (1982): 7–33.

Waterhouse, Benjamin. *The Botanist. Being the Botanical Part of a Course of Lectures on Natural History, Delivered in the University at Cambridge together with a Discourse on the Principle of Vitality.* Boston: Joseph T. Buckingham, 1811.

Waterman, Catherine H. *Flora's Lexicon: An Interpretation of the Language of Flowers.* Philadelphia: Hooker and Claxton, 1839.

Weems, Mason C. *The Life of Washington.* Edited by Marcus Cunliffe. Cambridge, Mass.: Belknap Press of Harvard University Press, 1982.

Welter, Barbara. "The Cult of True Womanhood: 1820–1860." *American Quarterly* 43 (1966): 151–74.

Whalen, Matthew D., and Mary F. Tobin. "Periodicals and the Popularization of Science in America, 1860–1910." *Journal of American Culture* 3 (1980): 195–203.

Whorton, James C. *Crusaders for Fitness: The History of American Health Reformers.* Princeton, N.J.: Princeton University Press, 1982.

Wight, Alex E. "Our New Courses in Botany." *Observer* 5 (1894): 332–33.

"Wild Flowers." *Youth's Companion* 19 (1849–50): 198–99.

Wilensky, Harold L. "The Professionalization of Everyone?" *American Journal of Sociology* 70 (1964): 137–58.

Willey, H. "Lichens under the Microscope." *American Naturalist* 4 (1870–71): 665–75.

Willis, Oliver R. *A Practical Flora for Schools and Colleges.* New York: American Book Co., 1894.

Wilson, Joan Hoff. "Dancing Dogs of the Colonial Period: Women Scientists." *Early American Literature* 7 (1973): 225–35.

Wood, Alphonso. *The American Botanist and Florist.* New York: A. S. Barnes and Co., 1870.

———. *A Class-Book of Botany.* Troy, N.Y.: Moore and Nims, 1846.

———. *Class-Book of Botany.* New York: A. S. Barnes and Burr, 1860.

———. *First Lessons in Botany.* Boston: Crocker and Brewster, 1843.

———. *Fourteen Weeks in Botany.* New York: A. S. Barnes and Co., 1879.

———. *Leaves and Flowers: or, Object Lessons in Botany.* A. S. Barnes and Co., 1860.

———. *Lessons in the Structure, Life, and Growth of Plants for Schools and Academies.* New York: A. S. Barnes and Co., 1889.

Wood, Ann Douglas. "'The Fashionable Diseases': Women's Complaints and Their Treatment in Nineteenth-Century America." *Journal of Interdisciplinary History* 4 (1973): 25–52.

Youmans, Eliza A. *Descriptive Botany.* New York: American Book Co., 1885.

———. *The First Book of Botany.* New York: D. Appleton and Co., 1873.

Zochert, Donald. "The Natural History of an American Pioneer: A Case
Study." *Transactions of the Wisconsin Academy of Sciences, Arts, and Letters*
60 (1972): 7–15.

———. "Science and the Common Man in Ante-Bellum America." *Isis* 65
(1974): 448–73.

Index

AAAS. *See* American Association for the Advancement of Science

Abbott, Jacob, 85–86, 93, 146

Academy of Natural Sciences (Philadelphia), 26–27

Agassiz Association, 140–45

Agriculture: and botany, 112–16, 121, 135–36

Alcott, Louisa May, 49, 79, 92–93, 146

Alexander, Fanny, 75

Amateurs (in botany): as experts, 2–3; changes in status of, 2–4, 98, 131–32; sources for study of, 7–8; early cooperation with professionals, 22–37; increasing split with professionals, 37, 97, 130–32; and natural theology, 108–9; differences between professionals and, 149–50. *See also* Botanical clearinghouses; Botanizers; Botanizing; Children; Hobby(ies); Information networks; Motivations; Nature-Study movement

American Association for the Advancement of Science (AAAS), 13, 24–25, 31–32, 124, 130

American Astronomical Society, 131–32

American Botanical Club, 130

American Journal of Education, 41, 59–60

American Journal of Science, 29–30, 35, 66, 129

American Medical Botany (Bigelow), 117

American Naturalist, 129–30

American Philosophical Society, 26

American Practice of Medicine (Beach), 117–18

Ames, Oakes, 96–97

Antioch College, 77

Appleton's, 129

"Argument from design," 101; God's character in, 105–6; Asa Gray's contribution to, 109

Artificial systems of classification. *See* Linnaean system of classification

Asa Gray Memorial Botanical Chapter (Agassiz Association), 142–43

Astronomy, 106; amateurs in, 3, 31, 131–32; in secondary-school curriculum, 54, 56; and religion, 104

Atlantic Monthly, 35, 129

Austin, Coe Finch, 18

Austin, Mrs. Coe Finch, 18

Austin, Rebecca, 35

Bache, A. D., 31

Bacon, Francis, 103

Bailey, Liberty Hyde, 135–36

Bailey, Margaret Emerson, 138

Bailey, William Whitman, 18, 138; on utility of botany, 7, 121–22; on field guides, 16–18

Baily, William L., 110

Baird, Spencer F., 33–34

Ballard, Harlan H., 140–41

Barnes, Kate, 126

Barton, Benjamin Smith, 14

Bartram, John, 26

Bartram, William, 11, 71
Beach, Wooster, 117–18
Bebb, Michael Schuck, 21
Bessey, Charles Edwin, 116,
130, 135; as New Botanist,
127, 128; efforts to introduce
New Botany into secondary-
school curriculum, 133, 136,
138
Bigelow, Jacob, 117
Biographical Dictionary of Ameri-
can Science, 13
Biology: emphasis on, in New
Botany, 36–37, 127, 133–34,
148
Bloomer outfits, 19, 77
Boston Globe, 2
Boswell, James, 71
Botanical Bulletin, 125
Botanical clearinghouses, 32–36
Botanical Club (of California), 80
Botanical Club of the AAAS, 130
Botanical education: changes in,
51–68; in nonbotany courses,
64; and natural theology, 100.
See also Botanical textbooks;
Classification systems; Colleges; Educators; Elementary
schools; Secondary-school curriculum
Botanical Gazette, 80, 128–29,
131; and botany as a hobby,
95; as professional journal,
124, 125–27
Botanical journals, 22–23, 24,
35; growth of, 27, 29–30, 124,
129–30
Botanical societies: links among,
22, 24; growth of, 26–27;
members of, 29; leadership of,
37; professionalization of, 124,
130–31
Botanical Society of America,
37, 124, 130–31
Botanical Textbook (Gray), 65,
67

Botanical textbooks: natural theology in, 100–102. *See also*
titles of specific textbooks
Botanizers (amateur botanists):
motivations of, 1; work of, 9–
21; number of, 11–12. *See*
also Amateurs; names of well-
known botanizers
Botanizing: equipment for, 12,
16; worklike aspects of, 83–
85; continued popularity of, in
spite of New Botany, 132–33,
134, 145. *See also* Amateurs;
Nature-Study movement
Botanizing (Bailey), 7, 18
Botany: popularity of, 1, 11; scientists' view of teaching of, 65;
differences in, as studied by
professionals and amateurs,
123; as process rather than
body of knowledge, 140; as
hobby, 146. *See also* Amateurs; Botanical education; Botanical journals; Botanical
societies; Botanizing; Natural
history; Nature-Study movement; New Botany; Profes-
sionals; Science and religion;
Taxonomy
Botany for Beginners (Lincoln
Phelps), 60–62, 88, 104
Botany for High Schools and Col-
leges (Bessey), 128, 133
Botany for Young People (Gray),
101, 104
Botany of the Southern States
(Darby), 113
"Botany's Charms," 95
Boys: botanizing clothing of, 18–
19. *See also* Men
Brandegee, Mary Katherine, 80–
81
Brandegee, Townshend Stith,
80–81
Bulletin of the Torrey Botanical
Club, 125–26, 129, 131

ence in America in the Nineteenth Century" (Reingold), 12
"Delta," 92
Describing specimens: work involved in, 85
Dickinson, Emily, 70
Dictionary of American Biography, 13
Discipline formation, 147
"Divine providence": doctrine of, 101–2, 106–7, 109
Dredges, 15
Drying (of plants). *See* Preservation
Dwight, Timothy, 41–42

Eastwood, Alice, 81
Eaton, Amos, 66; as popularizer of botany, 11; sales of textbooks by, 12; goals for botanical education, 52, 58–64, 90; differences with Gray, 65; and Linnaean classification system, 67; and women botanizers, 69, 71, 76; and natural theology, 100–101. See also *Manual of Botany for the Northern States*
Eclectic medicine, 117–18, 119, 121
Educators: interest in botanizing, 44–46, 47; goals in teaching botany, 51–52; and amateur botanists, 140–45. *See also* Botanical education; Botanical textbooks; Colleges; Classification systems; Elementary schools; Secondary-school curriculum
Eight Cousins (Alcott), 49, 79
Elementary schools: Nature-Study in, 136, 138–39
Elements of Botany (Gray), 65
Elements of Botany Addressed to a Lady (Rousseau), 70
Emerson, Ralph Waldo, 96

Empiricism: and natural theology, 103
Employment: as characteristic of a profession, 4–6; of collectors, 23–24, 35; and professionalization of botany, 30, 37
Entomology: collecting of specimens in, 44
Equipment: botanical, 15–21
Essais élémentaires sur la botanique (Rousseau), 70
Evolution, 109
Exchanging (of specimens): botanizers' interest in, 9, 23; breakdown in networks of, 124
Exercise: and collecting trips, 14; as benefit of botanizing, 38, 44, 46, 48, 50, 83, 85, 108, 148; and botanical education, 60; and botanizing for women, 74, 77, 78. *See also* Invalids
Expenses: of a botanizer, 12–13, 16, 149
Experimentation: emphasis on, in New Botany, 37, 127, 128, 129, 133, 149
Experts: amateurs as, 2–3; professionals as, 4–6, 30. *See also* Amateurs; Botanizers; Professionals

Familiar Lectures on Botany (Lincoln Phelps), 60–63, 65, 66, 101
Fern study, 2, 142–43
Fiction: as source of information about botanizing, 8; portrayal of botanizing in, 49, 78–79, 138–39, 146; for children, 85–88; natural theology in, 100
Field guides, 16; sales of botany, 12; amateurs' criteria for, 20. *See also* titles of specific field guides

Leisure. *See* Recreation
Lemmon, J. G., 80, 134; and botanical correspondence webs, 34–36
Lemmon, Mrs. J. G., 80
"Liberal Education for Farmers," 116
Lincoln Phelps, Almira Hart, 66; and botanical education, 52, 57–64, 90; differences between Gray and, 65; and Linnaean classification system, 67; collecting practices of students, 75–76, 78, 79; and natural theology, 101, 104, 108. *See also Botany for Beginners; Familiar Lectures on Botany*
Linnaean Fern Chapter (Agassiz Association), 142–43
Linnaean Society, 27
Linnaean system of classification, 10, 11, 52; use by amateurs for plant identification, 20, 30, 147; Eaton's and Lincoln Phelps's use of, 62–63; contrasted with other systems, 62–64; Gray's disdain for, 65
Linneaus, Carl, 52. *See also* Linnaean system of classification
Literacy: as necessity for botanizers, 12–13
Literary symbolism: and botany, 43
Logan, James, 26
Lowell, James Russell, 140
Lyceum movement: and botany, 11, 127–28

Manual of Botany for the Northern States (Eaton), 60–62, 65, 66; sales of, 12; natural theology in, 100–101
Manual of the Botany of the Northern United States (Gray), 34
Massachusetts State Assembly (of the Agassiz Association), 141–42
Mathematics: botany likened to, for mental discipline, 44–45, 50
Mead, S. B., 13
Medical Repository, 29
Medicine: and botany, 54, 64, 113, 116–19
Memory. *See* Mental discipline
Men: botanizing clothing of, 18–19; as collectors for women botanists, 19, 75–76; and botanizing, 70, 72, 74
Mental discipline: botany as an aid to, 38, 44–48, 83, 85, 108; and botanical education, 52, 60, 62. *See also* Observation
Middle-class values: in botanical pursuits, 13; in the Agassiz Association, 143–44. *See also* Natural theology; Piety; Self-improvement; Work ethic
Mills, John Stuart, 47
Mitchell, S. Weir, 121
Moral benefits: of studying botany, 7. *See also* Self-improvement
Morang, Thomas, 13
Motivations: differences between professional and amateur botanists, 3–4, 6–7, 94–96, 147, 148–49; of collectors, 24, 128–29. *See also* Exercise; Gentility; Mental discipline; Moral benefits; Natural theology; Personal advancement; Personal enrichment; Pleasure; Self-improvement; Social activity; Social reform; Utility; Work ethic
Mount Holyoke, 74, 76

National Education Association: Committee of Ten, 136

National Quarterly Review, 61
Natural history: as orientation of
    botanizers, 9–21, 148; trend
    away from, to biology, 36–37,
    127; defined, 56–57
Natural history circle, 25–26,
    32
Natural philosophy. *See* Physics
Natural system of classification:
    scientists' preference for, 30,
    64–68, 147–48; in botanical
    education, 52; described, 62–
    64. *See also* Linnaean system
    of classification
Natural theology, 101; and bota-
    nizing, 7, 43, 99–111, 146,
    148–49; and botanical educa-
    tion, 52, 60; botany as part of
    courses on, 64; and natural
    systems of classification, 67;
    defined, 99; as form of self-
    improvement, 103–6, 111; in
    Agassiz Association, 143–44.
    See also Piety
Nature-Study movement: educa-
    tors' support for, 52; observa-
    tional emphasis in, 68; and
    amateur botany, 132–33, 135–
    45; defined, 136; appeal of,
    145; professionals' lack of in-
    terest in, 149
Nature-Study Review, 139
New Botany, 135; laboratory em-
    phasis in, 37, 68, 149; as sub-
    ject for professionals, 52, 95,
    126, 127–34, 148–49; de-
    fined, 127; amateurs' lack of
    interest in, 128–29
*New Guide to Health or Botanic
    Family Physician* (Thomson),
    117
Newpark School (Lewisberry,
    Pa.), 89
New World: botanizing in, 9–10
New York Academy of Sciences,
    127–28

New York Lyceum of Natural His-
    tory, 127–28
North American Review, 30

Observation: as important part of
    botanizing, 37, 44, 45–46, 47;
    as part of botanical education,
    52, 60, 62. *See also* Mental
    discipline
Observer, 76, 95, 141, 142
Ornithology: amateurs in, 3; col-
    lection of specimens in, 44

Paley, William, 101
Pastime of Learning, The (anon.),
    89
Patapsco Female Institute (Md.):
    Lincoln Phelps's work at, 58–
    59, 74; collecting practices at,
    76, 79
Pedagogical issues in botany. *See*
    Botanical education; Classi-
    fication systems; Colleges; Ele-
    mentary schools; Secondary-
    school curriculum
Perfectionism, 39
Personal advancement, 108; as
    motivation for botanizers, 38–
    39, 44, 46–50
Personal enrichment: as motiva-
    tion for botanizers, 4, 6, 95,
    148–49; as goal of Nature-
    Study movement, 145
Pestalozzi, Johann Heinrich, 59–
    60
Phelps, Almira Hart Lincoln. *See*
    Lincoln Phelps, Almira Hart
Phelps, Elizabeth Stuart, 49
Physical culture. *See* Exercise
Physicians: and botany, 119. *See
    also* Medicine
Physics: in secondary-school cur-
    riculum, 54, 56
Piety: and botanizing, 43, 103–
    6, 108. *See also* Natural theol-
    ogy

Plant pathology, 116
Plant presses, 20–21
Plant World, 95
Pleasure: as motivation for botanizers, 24, 90–94; lack of, in studying New Botany, 133–34
Plummer, Sarah, 35
Poland Springs Water Cure, 48
Popular Science Monthly, 129, 130
Popular Science News, 141
Potter, Henry, 2–3
Practicality. *See* Utility
"Practitioners": Reingold's definition of, 6
Preservation: botanizers' interest in, 9; methods of, 20–21; work involved in, 85
Pringle, Cyrus Guernsey, 2, 3
Professionalization: of botany, 1, 123–34, 146–49; of American scientific community, 3–7, 36–37. *See also* Professionals
Professionals: role in scientific community, 3; changing definitions of, 4–5; perception of botany by, 94–96; decreasing reliance on amateurs by, 97; and natural theology, 100, 107–8. *See also* Amateurs; New Botany; Professionalization
Protestant theology: in botanizing, 13. *See also* Natural theology
Providence (R.I.) Franklin Society, 27, 124

Readers (schoolbooks): natural theology in, 100
Reasoning. *See* Mental discipline
Recreation: botanizing as, 83–98
Reform clothing. *See* Bloomer outfits
Reingold, Nathan, 5–6, 12, 132
Religion. *See* Natural theology

Rensselaer, Stephen Van, 59
Rensselaer Institute (Troy, N.Y.), 58, 59, 60, 63, 75, 76
Research. *See* Experimentation
"Researchers": Reingold's definition of, 6; estimated numbers of, 12
Reverchon, J., 80
Roessler, Mrs. S. E., 76
Rogers, Annie F., 80
Rollo's Museum (Abbott), 85–86
Rothenberg, Marc, 132
Rousseau, Jean-Jacques, 70
Royal Society of London, 25
Rust, Mary O., 126

St. Nicholas Magazine, 92, 141–42
Sandburg, Carl, 137–38
Santa Claus, 141
Schools. *See* Botanical education; Children
Science: advancement of, 3, 6; distinction between promotion and advancement of, 130; professionals define, 132. *See also* Botanizing; Botany; New Botany; Science and religion
Science, 70, 129, 130
Science and religion: 19th-century changes in relations between, 99–111. *See also* Natural theology
Scouting: Nature-Study in, 144
"Seaweed Album, The" ("Delta"), 92
Secondary-school curriculum: and early botanical education, 53, 54; efforts to teach New Botany in, 129, 133, 136–37; Nature-Study as part of, 137–38
Self-improvement: as motivation for botanizers, 7, 13, 24, 38–50, 146; economic component of, 39; religious basis of, 39.